T0269116

Moving Particle Semi-implicit Method

MOVING PARTICLE SEMI-IMPLICIT METHOD
Recent Developments and Applications

GEN LI
South China University of Technology, Guangzhou, China

GUANGTAO DUAN
The University of Tokyo, Tokyo, Japan

XIAOXING LIU
Sun Yat-Sen University, Guangzhou, China

ZIDI WANG
Japan Atomic Energy Agency, Japan

ELSEVIER

Elsevier
Radarweg 29, PO Box 211, 1000 AE Amsterdam, Netherlands
The Boulevard, Langford Lane, Kidlington, Oxford OX5 1GB, United Kingdom
50 Hampshire Street, 5th Floor, Cambridge, MA 02139, United States

Notices
Knowledge and best practice in this field are constantly changing. As new research and
experience broaden our understanding, changes in research methods, professional practices,
or medical treatment may become necessary.

Practitioners and researchers must always rely on their own experience and knowledge in
evaluating and using any information, methods, compounds, or experiments described
herein. In using such information or methods they should be mindful of their own safety
and the safety of others, including parties for whom they have a professional responsibility.

To the fullest extent of the law, neither the Publisher nor the authors, contributors, or
editors, assume any liability for any injury and/or damage to persons or property as a matter
of products liability, negligence or otherwise, or from any use or operation of any methods,
products, instructions, or ideas contained in the material herein.

ISBN: 978-0-443-13508-8

For Information on all Elsevier publications
visit our website at https://www.elsevier.com/books-and-journals

Publisher: Megan Ball
Editorial Project Manager: Sara Valentino
Production Project Manager: Fizza Fathima
Cover Designer: Christian J. Bilbow

Typeset by MPS Limited, Chennai, India

Contents

About the authors

Gen Li is a professor at South China University of Technology. He obtained his PhD degree from Waseda University, Japan, in 2015, and then joined the faculty of Xi'an Jiaotong University, China. In 2021, he moved to South China University of Technology. Prof. Li has been working on the particle method development and the method application in nuclear engineering for many years. He proposed an axisymmetric multiphase moving particle semi-implicit (MPS) method and applied the MPS method to the simulations of nuclear reactor thermal hydraulics and severe accident phenomena. He has published more than 30 journal papers and received two projects from the National Natural Science Foundation of China.

Guangtao Duan is an assistant professor at The University of Tokyo. He got his PhD degree from Xi'an Jiaotong University, China, in 2016. He worked as a postdoctoral researcher in the University of Tokyo (2016−19) and Waseda University (2017−19). From 2020, he got his current position at the University of Tokyo. Dr. Duan has made excellent contributions to accuracy improvement, stability enhancement, and multiphase-flow applications in particle methods, such as a high-accuracy particle method, a novel stable free-surface-detection method, sharp interface algorithms in multiphase flow, and advanced solidification and evaporation models. He has published more than 30 papers on international journals. Dr. Duan is a project principal investigator of the Japan Society for the Promotion of Science.

Xiaoxing Liu is an associate professor at Sun Yat-Sen University. He completed his PhD study in Kyushu University, Japan, in 2012−15. After that, he worked as a postdoctoral research fellow in the same laboratory for his PhD study. In 2020, he joined the faculty of Sun Yat-Sen University. His research interests focus on the meshless particle methods and simulations of nuclear reactor severe accidents. Dr. Liu also published many research papers related to advanced numerical models in particle method. He has served as a session chair in many international conferences (IACM, NUTHOS, etc.).

Zidi Wang is a researcher at the Nuclear Safety Research Center, Japan Atomic Energy Agency. He received his BSc degree (2012) and MSc degree (2015) from Xi'an Jiaotong University and Shanghai Jiao Tong University, respectively. After that, he received his PhD (2018) from the University of Tokyo, Japan. His research interests include the development of particle method for free-surface/interface flows and its applications to nuclear safety analysis.

CHAPTER ONE

Introduction

The numerical methods of fluid dynamics can be classified as grid method and particle method, according to the discretization scheme of the computation domain. The basic idea of grid method is grid division and forming the global discrete equations by the unit stiffness matrix located at discrete nodes or elements. Then, given the initial values and boundary condition the governing equations are solved. The grids are fixed at their place and do not move as the fluid flows, which is named as Eulerian description. In comparison the particle method discretizes the computation domain into particles, without any grid. Particle method solves the governing equations through the particle interactions. One particle interacts with its neighboring particles that have distances less than a certain radius. The particles move with their velocities, which is expressed as Lagrangian description.

Free surface and interface capture is an important aspect of computational fluid dynamics. When the computational domain is fixed or deformation is small the grid method can provide an accurate result. When dealing with the problems involving a large deformation (e.g., free surface, fragmentation, and coalescence), particle methods can break the restriction of fixed topology among the nodes and successfully avoid the complicated process in grid method, that is, mesh distortion and the conversion of old and new grids. The following two subsections will briefly introduce the conventional grid-based numerical method and the advanced meshless particle method.

1.1 Grid-based numerical method

In computational fluid dynamics the problems of large deformation and distortion, free surface capture, and those cases involving strong discontinuity in the physical field have been treated as numerical challenges. Several interface capturing methods in Eulerian or Euler-Lagrangian manner can be found in the literature, such as volume-of-fluid (VOF), level set, and front-tracking methods.

Moving Particle Semi-implicit Method.
DOI: https://doi.org/10.1016/B978-0-443-13508-8.00001-9
1

The VOF method uses a scalar quantity that represents the volume fraction of the fluid in each stationary mesh to identify the meshes located at the fluid interface. The position of the interface is updated by solving the VOF advection equation with the finite volume scheme to ensure the mass conservation. However, it lacks accuracy in direct calculation of normal vector and curvature because of the discontinuous spatial derivatives of the volume fraction around the interface [1,2]. Moreover the VOF method requires a sophisticated algorithm to reconstruct the interface and to avoid diffusion across the interface. It is difficult to accurately construct the local interface from the volume fractions because of numerical diffusion arising from fixed-point interpolations of advection terms in the VOF-function transport equation [3,4]. Though the VOF method has been widely used in two dimensions, it encounters a great challenge when applied in three dimensions.

The level-set method is also an Eulerian-based method, where the interface is defined as a zero level set of a distance function from the interface. In contrast with VOF method, this approach is straightforward to implement for both two and three dimensions. In level-set method the interface is descripted by a level-set function, and then the normal vector and curvature can be calculated by a continuous smoothed distance function [2]. Though the interface can be traced by solving the distance function, the mass is difficult to be conserved at the interface due to the sensitivity of numerical dissipation [5]. Some complex procedures to reinitialize and advect the level set function are needed to guarantee mass conservation.

In front-tracking method, based on Eulerian—Lagrangian manner the flow field is solved in a fixed grid and the position of the interface is tracked by a set of front marks. These markers can be classified as separated or connected. In practical applications, there are two common front-tracking techniques. One is representing the front interface by a group of moving marks and solving the entire flow field on background grids, where the grids are modified only near the front to make the grids line follow the interface. The other is representing the interface by a crowd of connected marks that carry forces. Then the interface forces are interpolated onto the grids to solve the "one-fluid" formulation of flow equation, where the background grids are fixed even near the front interface. These approaches can acquire a sharp interface, but it is complicated to implement and cannot maintain communication between interface grids and stationary grids [6,7].

The grid-based Eulerian methods using the stationary mesh have difficulties in maintaining interface sharpness, while those using the deformable mesh have difficulties in mesh adaptation to follow the highly
deformed free surfaces and complex interfaces. Numerical dissipation and
mesh distortion are the common problems in grid-based methods. In contrast, Lagrangian approaches calculate the convective term through the
motion of particles, which are capable of avoiding the numerical dissipation inherently. Moreover, these methods explicitly portray the motion
and deformation of the interface through the motion of particles in different phases without introducing extra interface transport equations.
Therefore Lagrangian methods have great potential for simulating the
multiphase and surface flows.

1.2 Particle-based numerical method

During the past two decades, Lagrangian meshless particle method
as a new generation computational fluid dynamics method has attracted
significant attention. Particle method simulates fluids and solids by discretizing the computation domain into particles that carry the information of
physical properties, velocity, and pressure. The grid generation is not necessary. The particles are distributed only in the area where the fluid presents so that even complex moving boundaries and fluid interfaces can be
readily captured. Particle motion is implemented by solving particle interaction models. Particle method has remarkable advantages in capturing
fluid-free surfaces and interfaces in the violent fluid flows with large
deformation and multiphase flows involving fluid breakup and coalescence. Moreover, numerical diffusion arising from the discretization of
advection term in the Navier—Stokes equation is automatically avoided
due to the adoption of Lagrangian description.

Two prevailing particle methods are the smooth particle hydrodynamics (SPH) method [8,9] and the moving particle semi-implicit (MPS)
method [10]. SPH is a method based on the integral representation of
quantities and spatial derivatives, which is originally developed using a
fully explicit algorithm for compressible nonviscous flows. It finds wide
applications in astrophysics, where the density changes so much ranging
from the inside of stars to vacuum, but the application of the method is

later expanded to solid and fluid mechanics as well. By contrast, MPS method is based on Taylor series expansion, which is originally developed for incompressible fluid flow using a semi-implicit algorithm. To date, however, as many advancements have been made to particle method, SPH and MPS methods have similar features in fluid mechanics. For example a pressure Poisson equation (PPE) that is solved using an implicit algorithm is introduced to SPH method [11], and a fully explicit algorithm is also applied to MPS method by replacing PPE with the equation of state [12−14]. They can both be used for the simulations of compressible and incompressible flows, respectively, with the relevant improvements, such as ISPH [11,15−17] and WC-MPS [12,14]. Nevertheless, the spatial discretization formulations of the SPH and MPS methods have significant differences.

Since MPS method has been brought forward for about two decades, it is considered as a promising but immature method. Some inherent drawbacks still exist, such as low accuracy, numerical instability, and physical inconsistency. The original discretization schemes of SPH and MPS methods only have the zeroth-order accuracy, and the numerical instability occurs easily due to the nonuniform distribution of particles. In particle method, moreover, the neighboring particles are memorized in a neighbor list and it is updated in every time step. This is an additional procedure in each time step and time-consuming in contrast to mesh generation that is carried out once in the initialization process. Many studies have been working on the method development to make them more capable of simulating compressible and incompressible multiphase flows in nuclear engineering and ocean engineering.

References

[1] L. Qian, Y. Wei, F. Xiao, Coupled THINC and level set method: a conservative interface capturing scheme with high-order surface representations, J. Comput. Phys. 373 (2018) 284−303.
[2] Z. Wang, J. Yang, F. Stern, A new volume-of-fluid method with a constructed distance function on general structured grids, J. Comput. Phys. 231 (2012) 3703−3722.
[3] V.R. Gopala, B.G.M. van Wachem, Volume of fluid methods for immiscible-fluid and free-surface flows, Chem. Eng. J. 141 (2008) 204−221.
[4] M. Sussman, E.G. Puckett, A coupled level set and volume-of-fluid method for computing 3D and axisymmetric incompressible two-phase flows, J. Comput. Phys. 162 (2000) 301−337.
[5] J.A. Sethian, P. Smereka, Level set methods for fluid interfaces, Annu. Rev. Fluid Mech. 35 (2003) 341−372.
[6] J. Hua, J.F. Stene, P. Ling, Numerical simulation of 3D bubbles rising in viscous liquids using a front tracking method, J. Comput. Phys. 227 (2008) 3358−3382.

[7] S.O. Unverdi, G. Tryggvason, A front-tracking method for viscous, incompressible, multi-fluid flows, J. Comput. Phys. 100 (1992) 25–37.

[8] L.B. Lucy, A numerical approach to the testing of the fission hypothesis, Astronomical J. 82 (1977) 1013–1024.

[9] R.A. Gingold, J.J. Monaghan, Smoothed particle hydrodynamics: theory and application to non-spherical stars, Mon. Not. R. Astron. Soc. 181 (1977) 375–389.

[10] S. Koshizuka, Y. Oka, Moving particle semi-implicit method for fragmentation of incompressible fluid, Nucl. Sci. Eng. 123 (1996) 421–434.

[11] S.J. Cummins, M. Rudman, An SPH projection method, J. Comput. Phys. 152 (1999) 584–607.

[12] A. Shakibaeinia, Y.C. Jin, A weakly compressible MPS method for modeling of open-boundary free-surface flow, Int. J. Numer. Methods Fluids 63 (10) (2010) 1208–1232.

[13] A. Tayebi, Y.-C. Jin, Development of Moving Particle Explicit (MPE) method for incompressible flows, Computers & Fluids 117 (2015) 1–10.

[14] M. Jandaghian, A. Shakibaeinia, An enhanced weakly-compressible MPS method for free-surface flows, Computer Methods Appl. Mech. Eng. 360 (2020) 112771.

[15] R. Xu, P. Stansby, D. Laurence, Accuracy and stability in incompressible SPH (ISPH) based on the projection method and a new approach, J. Comput. Phys. 228 (18) (2009) 6703–6725.

[16] P. Nair, T. Poeschel, Dynamic capillary phenomena using Incompressible SPH, Chem. Eng. Sci. 176 (2018) 192–204.

[17] A. Khayyer, H. Gotoh, H. Falahaty, Y. Shimizu, An enhanced ISPH–SPH coupled method for simulation of incompressible fluid-elastic structure interactions, Computer Phys. Commun. 232 (2018) 139–164.

CHAPTER TWO

Original moving particle semi-implicit method

The basic idea of moving particle semi-implicit (MPS) is to discretize the fluid into many particles, which carry the position, pressure, and velocity information. Then the governing equations are discretized into the so-called particle interaction models, including the gradient, divergence, and Laplacian models. Next, based on the forces calculated from particle interaction models, the particles move in a Lagrangian framework, representing the flow. In this chapter the components of the original MPS method are described.

2.1 Governing equations

The governing equations for incompressible flow in the Lagrangian framework are used in the MPS method. Specifically the mass and momentum conservation equations are

$$\frac{D\rho}{Dt} = -\rho\nabla\cdot\mathbf{u} = 0, \tag{2.1}$$

and

$$\frac{D\mathbf{u}}{Dt} = -\frac{1}{\rho}\nabla p + \frac{\mu}{\rho}\nabla^2\mathbf{u} + \mathbf{g}, \tag{2.2}$$

where $\mathbf{u}$ is the velocity, p is the pressure, $\mathbf{g}$ is the gravity, ρ is the density, and μ is the dynamic viscosity. To make the discussion simple, the subscale stress due to turbulent effect and the continuum forces due to surface tension are not considered in this chapter.

2.2 Basic particle interaction models

Before the particle interaction models are described, the weight function and the particle number density (PND) are presented first.

Moving Particle Semi-implicit Method.
DOI: https://doi.org/10.1016/B978-0-443-13508-8.00002-0

The weight function is used to measure the relative weight for the interaction forces among particles. The weight function only depends on the distance between two particles. Briefly the closer the two particles are, the stronger their interactions will be. Compared to smooth particle hydrodynamics (SPH), the restrictions to the weight functions in MPS are much less. For example, any function that is always positive and monotonically decreasing with particle distance can be used as the weight function in MPS. The most widely adopted weight function in MPS is

$$
w_{ij} = w\left(r_{ij}, r_e\right) = \begin{cases} \dfrac{r_e}{r_{ij}} - 1 & \text{if } r_{ij} < r_e \\ 0 & \text{if } r_{ij} \geq r_e \end{cases}, \tag{2.3}
$$

where r_{ij} is the particle distance between the reference particle i and a neighboring particle j, and r_e is the effective particle-interaction radius. Let l_0 denote the initial constant particle distance or the average particle distance. The effective radius is usually selected as $r_e = k \cdot l_0$. In most flow simulations, $k = 3.1$ would give a good balance between the computational cost and the discretization accuracy. A sketch for the weight function is shown in Fig. 2.1. It is noted that the weight function in MPS is dimensionless.

Based on the weight function, it is possible to define the PND for each particle. Specifically the dimensionless PND of the reference particle i is defined as

$$
n_i = \sum_{j \neq i} w\left(r_{ij}, r_e\right), \tag{2.4}
$$

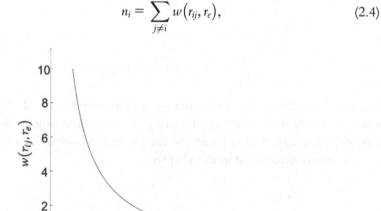

Figure 2.1 Sketch for the weight function in moving particle semi-implicit.

where $j \neq i$ indicates that the neighboring particles j do not include the reference particle i. When the particles are aggregated, the local PND will increase. While the particles are scattered, the local PND will decrease. Therefore the PND is approximately proportional to the fluid density, as discussed by Koshizuka and Oka [1]. The absolute value of PND is not important, and only the relative value of PND is meaningful. In this situation, a reference value for PND is necessary. Based on the initial regular particle distribution a constant PND, n^0, can be calculated for the internal particles. Thus n^0 is selected as the reference PND. Based on the notion of PND the mass conservation equation is equivalent to keeping the PND approximately constant around n^0 in the MPS simulations.

2.2.1 Gradient model

The gradient model is usually used to calculate the pressure gradient forces in MPS. Taking the pressure gradient as an example the gradient model is calculated from the weighted summation of the pressure gradient forces in different directions. The pressure gradient force between two particles is $(p_j - p_i)(\mathbf{r}_j - \mathbf{r}_i)/(r_{ij}^2)$, where p_i is the pressure of particle i, $\mathbf{r}_i$ is the position vector of particle i, and the particle distance can be calculated from $r_{ij} = \|\mathbf{r}_j - \mathbf{r}_i\|$. This force always drives the particle to move from the high-pressure region to the low-pressure region, which is consistent with the role of pressure in fluid dynamics. For a reference particle i, each neighboring particle j of it will have a relative weight as w_{ij}/n^0. Based on this relative weight the summation of the pressure-gradient forces in different directions will give the gradient model:

$$\langle \nabla p \rangle_i = \frac{d}{n^0} \sum_{j \neq i} \frac{p_j - p_i}{r_{ij}} \frac{(\mathbf{r}_j - \mathbf{r}_i)}{r_{ij}} w(r_{ij}, r_e), \qquad (2.5)$$

where d is the dimensional number. A sketch for the gradient model is presented in Fig. 2.2. The calculated gradient based on the weighted summation always points to the direction where the pressure increases fastest, thus driving the particle to move in the opposite direction due to the minus sign in the momentum-conservation equation. When the discretized summation above is converted to the continuous integral, the gradient model can also be derived mathematically, as derived by Isshiki [2]. When the integral version is considered, the dimensional number d will emerge as an integral coefficient.

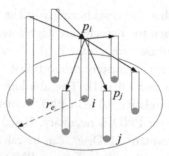

Figure 2.2 Sketch for the gradient model.

However, when the above gradient model is directly used for pressure gradient in the MPS method, instability can take place easily. This happens because the recovery capability of the gradient model to fluctuations is quite poor. For example, slight fluctuation will finally result in instability. To avoid the instability the following model was proposed by Koshizuka and Oka [1]:

$$\langle \nabla p \rangle_i = \frac{d}{n^0} \sum_{j \neq i} \frac{p_j - p_{i,min}}{r_{ij}} \frac{(\mathbf{r}_j - \mathbf{r}_i)}{r_{ij}} w(r_{ij}, r_e), \qquad (2.6)$$

where $p_{i,min}$ is the minimal pressure among the neighboring particle j and the reference particle i. It is noted that $p_{i,min}$ is calculated for each reference particle i separately. Because $p_j \geq p_{i,min}$, repulsive pressure-gradient forces are always guaranteed between any particle pair. These repulsive forces effectively prevent the continuous approach of two particles. Thus the simulation stability is greatly enhanced, even though the accuracy of the gradient model is slightly reduced. To maintain the stability near free surfaces, one special treatment must be adopted [3]: The negative pressure under the free surface must be reset to zero. This special treatment can effectively improve the stability at free surfaces. However, it makes the original MPS method not capable of dealing with negative pressures under free surfaces. In other words the original MPS method is only capable of simulating the problems with positive pressure under free surfaces.

Based on the same treatment for the pressure under free surfaces, another stable pressure gradient model is the conservative one [4]:

$$\langle \nabla p \rangle_i = \frac{d}{n^0} \sum_{j \neq i} \frac{p_j + p_i}{r_{ij}} \frac{(\mathbf{r}_j - \mathbf{r}_i)}{r_{ij}} w(r_{ij}, r_e). \qquad (2.7)$$

This model is essentially like the conservative model in SPH when the PND of each particle is close to n^0. However, when the negative pressure takes place, the above conservative model will produce the attractive forces and easily trigger instability. Thus the conservative model is not applicable to the physically negative pressure. Besides, the conservative model usually produces more stable results than the model in Eq. (2.6) due to the stronger numerical dissipation. In contrast, Eq. (2.6) could provide more accurate simulations than Eq. (2.7), especially for the multiphase flow simulations.

2.2.2 Divergence model

The divergence model is like the gradient model. The velocity divergence is taken as an example to explain how the divergence is obtained. As shown in Fig. 2.3 the linear strain rate between a pair of particles i and j is $\left(\mathbf{u}_j - \mathbf{u}_i\right) \cdot \left(\mathbf{r}_j - \mathbf{r}_i\right)/r_{ij}^2$, where $\mathbf{u}_i$ is the velocity of particle i. When this value is positive the distance between particle i and j will increase. Thus this value measures how fast a particle j is leaving particle i. The summation of the linear strain rate for all the neighboring j based on a dimensionless weight w_{ij}/n^0 will give the divergence model:

$$\langle \nabla \cdot \mathbf{u} \rangle_i = \frac{d}{n^0} \sum_{j \neq i} \frac{\mathbf{u}_j - \mathbf{u}_i}{r_{ij}} \frac{\left(\mathbf{r}_j - \mathbf{r}_i\right)}{r_{ij}} w\left(r_{ij}, r_e\right). \tag{2.8}$$

The velocity divergence measures the rate of expansion of the fluid per unit volume. For example, when the velocity divergence is positive the fluid is in an expansion state. On the other hand, when the velocity divergence is negative the fluid is in a compression state. The above divergence model is consistent with the expansion and compression states due to the velocity distribution. Like the gradient model, the divergence model can also be derived based on the continuous integral of the linear strain rate in all the directions. In this situation the dimensional number d will emerge as an integral coefficient.

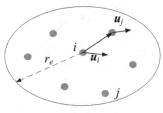

Figure 2.3 Sketch for the divergence model.

2.2.3 Laplacian model

The Laplacian model is derived based on the diffusion theory [1,5]. Basically the diffusion theory implies that each particle will diffuse the physical quantity it carries to the neighboring particles and receive the physical quantity diffused by the neighboring particles. Specifically, as shown in Fig. 2.4, a reference particle i carries the variable ϕ_i, which diffuses its quantity to the neighboring particles based on the weight function. The portion of ϕ_i transferred from particle i to particle j is $\phi_i w(r_{ij}, r_e)/n^0$. Meanwhile the portion of ϕ_j transferred from particle j to particle i is $\phi_j w(r_{ij}, r_e)/n^0$. Therefore the net quantity transferred between particles i and j is $(\phi_j - \phi_i)w(r_{ij}, r_e)/n^0$.

When the net quantity transferred between reference particle i and each neighboring particle j is summed up and normalized by a dimensional coefficient; the Laplacian model can be obtained as follows:

$$\langle \nabla^2 \phi \rangle_i = \frac{2d}{n^0 \lambda} \sum_{j \neq i} \left(\phi_j - \phi_i \right) w\left(r_{ij}, r_e\right), \tag{2.9}$$

where the variance of particle distance λ is defined as

$$\lambda = \frac{\sum_{j \neq i} r_{ij}^2 w\left(r_{ij}, r_e\right)}{\sum_{j \neq i} w\left(r_{ij}, r_e\right)}. \tag{2.10}$$

It is noted that λ is a constant in the simulation and it is calculated based on the initial regular particle distribution. The strict derivation of the Laplacian model based on the diffusion theory can be found in Ref. [5]. Meanwhile the mathematical derivation of the Laplacian based on the integral in the effective neighboring area can be found in Refs. [2,6]. When the Laplacian model is used to calculate the diffusion of velocity,

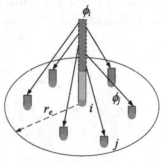

Figure 2.4 Sketch for the Laplacian model.

the velocity portion diffused by a particle is just received by a neighboring particle. Thus the momentum conservation is exactly maintained. This also holds for the heat transfer.

Briefly, all the gradient, divergence, and Laplacian models in MPS exhibit clear physical meaning. Meanwhile, they can be derived mathematically. These features make the MPS method produce physically consistent and reasonable results.

2.3 Boundary conditions

The complete neighbor support plays a key role in maintaining the accuracy of the MPS models. Therefore the variables of the neighboring boundary particles must be carefully considered. Meanwhile the physical boundaries are usually considered in MPS.

2.3.1 Free surface

In the original MPS method a free surface boundary is detected from the decrease of PND. Specifically the following condition is satisfied:

$$n_i < \beta \cdot n^0, \qquad (2.11)$$

where β is a detection coefficient; particle i will be detected as a free-surface particle. The typical value is $\beta = 0.97$ [1]. When β is large, more internal particles will be detected as free-surface particles, thus reducing the accuracy. On the other hand, when β is small, fewer internal particles are detected as free-surface particles, easily generating fluctuations and even triggering instability. The detected free-surface particles always take a constant zero pressure. In this manner the free-surface boundary is imposed.

2.3.2 Wall boundary

The imposition of the wall boundary conditions is sketched in Fig. 2.5, where one layer of wall particles representing the exact wall boundary and three layers of dummy particles compensating the PND of wall particles are used. From the viewpoint of velocity the wall boundary can be clarified into the no-slip and free-slip boundaries. When the no-slip boundary is considered the wall particles take zero velocity, and the dummy particles take the opposite velocity to the velocity of the mirror points with respect to the wall boundary, as shown in Fig. 2.5A.

However, it must be noted that this method would be difficult for the wall boundary with complicated geometric shapes. When the free-slip boundary is considered the dummy and wall particles are assumed to take the same velocity with reference particle i, as shown in Fig. 2.5B. In this manner, there is no relative velocity between the reference particle and the wall/dummy particles, indicating the free-slip wall boundary [7].

The pressure boundary condition at the wall is imposed as follows: A Neumann boundary condition with zero flux is considered for pressure. The treatment of the pressure of the dummy particles is similar to the free-slip velocity boundary condition. Specifically the pressure of wall particles is solved from the pressure Poission equation (PPE) like fluid particles. It is always assumed that the pressure of dummy particles is the same with the reference pressure, namely, $p_j = p_i$, for dummy particle j. In this situation, when the PPE is solved the dummy particles can be directly neglected due to the zero flux. When the pressure gradient is calculated by Eq. (2.6) the pressure difference $p_j - p_{i,min}$ is simplified to $p_i - p_{i,min}$ for the dummy neighboring particle j. When the pressure gradient is calculated by Eq. (2.7) the pressure summation $p_j + p_i$ is simplified to $p_i + p_i$ for the dummy neighboring particle j. For example the dummy particles can still contribute to the stable pressure gradient models. This contribution can effectively prevent the fluid particles penetrating into the wall boundary.

2.3.3 Inflow boundary condition

The inflow and outflow boundaries are more difficult in the Lagrangian particle methods than in the Eulerian mesh methods. Shakibaeinia and Jin

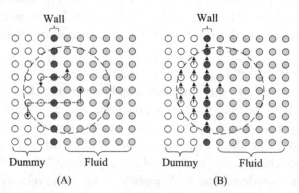

Figure 2.5 Sketch showing the treatment of no-slip and free-slip wall boundaries. (A) No-slip wall boundary. (B) Free-slip wall boundary [7].

[8] proposed to use the particle recycle to implement the inflow and outflow boundary conditions. Specifically the particles flowing out of the considered domain will be rearranged at the inflow boundary.

The configuration of the inflow boundary is shown in Fig. 2.6. The roles of the inflow and inflow-dummy particles are like those of the wall and dummy particles in Fig. 2.5 in terms of pressure calculation, respectively. For example, the inflow particles will be considered when the PPE is solved. The inflow-dummy particles do not participate the PPE solution, and they are mainly used to compensate the PND decrease near the inflow boundary. The inflow and inflow-dummy particles are prescribed with the given velocity and always move with the given velocity. When the inflow-dummy particles move across Line 2 in Fig. 2.6, they will be converted into the inflow particles. At the same time the same number of additional inflow-dummy particles will be added at the left side of the inlet to supplement the inflow-dummy particles. Similarly, the inflow particles will be converted into the normal fluid particles when they move across Line 1 in Fig. 2.6. The advantage of this inflow arrangement is that it can suppress the inflow fluctuations as much as possible [7]. The outflow boundary in this manner would be much difficult because the outflow velocity is usually unknown. In this situation, an easy way is to set the outflow boundary directly to the free surface boundary. For example, there is no need to arrange additional particles out of the outflow line. Particularly, when the fluid particles move across the outflow line, they will be removed and stored for recycling.

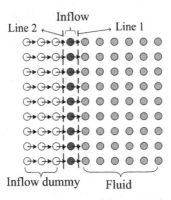

Figure 2.6 Sketch showing the treatment of inflow boundary condition [7].

2.4 Time marching algorithm

2.4.1 Semi-implicit algorithm

The prediction—correction algorithm is usually adopted in MPS, where the pressure can be calculated implicitly or explicitly. When pressure is calculated implicitly, the algorithm becomes the semi-implicit algorithm.

1. The viscosity and gravity (external force) terms in the governing Eq. (2.2) are calculated explicitly, and a temporary velocity $\mathbf{u}_i^*$ is obtained:

$$\mathbf{u}_i^* = \mathbf{u}_i^k + \left(\frac{\mu}{\rho}\langle\nabla^2\mathbf{u}\rangle_i^k + \mathbf{g}\right)\Delta t. \tag{2.12}$$

2. The convection is calculated by the particle motion based on $\mathbf{u}_i^*$, and a temporary position $\mathbf{r}_i^*$ is obtained:

$$\mathbf{r}_i^* = \mathbf{r}_i^k + \mathbf{u}_i^*\Delta t. \tag{2.13}$$

Based on $\mathbf{r}_i^*$ a temporary PND n_i^* is calculated. The pressure gradient should be considered next to keep the mass conservation. The velocity change produced by the pressure gradient is

$$\Delta\mathbf{u}_i' = -\frac{\Delta t}{\rho}\langle\nabla p\rangle_i^{k+1}. \tag{2.14}$$

This velocity change should just compensate the derivation of the PND based on the mass conservation equation, (Eq. 2.1). For example, one has

$$\nabla\cdot\Delta\mathbf{u}_i' = -\frac{1}{\Delta t}\frac{n_i^{k+1} - n_i^*}{n^0}, \tag{2.15}$$

where the PND at the next time is expected to be restored to n^0, namely, $n_i^{k+1} \to n^0$. Taking the divergence of both sides of Eq. (2.14) will finally give the PPE (Eq. 2.16) in MPS.

3. The pressure is implicitly solved from the derived PPE as follows:

$$\frac{1}{\rho}\langle\nabla^2 p\rangle_i^{k+1} = -\frac{1}{\Delta t^2}\frac{n_i^* - n^0}{n^0}, \tag{2.16}$$

4. The final velocity at the next time step is updated by considering the pressure-gradient velocity change:

$$\mathbf{u}_i^{k+1} = \mathbf{u}_i^* + \Delta\mathbf{u}'_i. \tag{2.17}$$

5. The final position is updated as follows:

$$\mathbf{r}_i^{k+1} = \mathbf{r}_i^* + \Delta\mathbf{u}'_i \Delta t. \tag{2.18}$$

The summary of the solution algorithm above [9] is presented in Fig. 2.7. It is noted that the pressure gradient model in Eq. (2.17) is usually discretized by Eq. (2.6), even though Eq. (2.7) can also provide stable results.

2.4.2 Explicit algorithm

The key difference between the explicit and implicit algorithms is how to calculate the pressure. In the explicit algorithm [8] the pressure is directly calculated from the PND:

$$p_i^{k+1} = \rho_0 c_0^2 \left(\frac{n_i^* - n^0}{n^0} \right), \tag{2.19}$$

where ρ_0 is the fluid density and c_0 is the artificial sound speed (usually $c_0 = 10 u_{max}$, where u_{max} is the maximum velocity in the flow field). When the pressure calculation in Eq. (2.19) replaces the pressure solution in Eq. (2.16), the implicit algorithm above becomes the explicit algorithm. It must be noted that Eq. (2.7) is usually used as the pressure gradient model in the explicit algorithm because this model can provide better stability. Another difference between the implicit and explicit algorithms is the restriction on time steps, as discussed below.

2.4.3 Restrictions on time step

In the semi-implicit algorithm the time step is usually restricted by the CFL condition due to convection as follows [1]:

$$\Delta t < 0.15 \frac{l_0}{u_{max}}, \tag{2.20}$$

where u_{max} is the maximum velocity in the flow field.

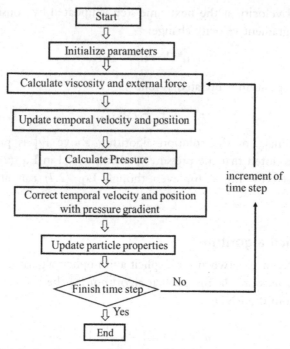

Figure 2.7 Sketch showing the solution algorithm in moving particle semi-implicit [9].

In the explicit algorithm the time step is usually restricted by the CFL condition due to the sound speed as follows:

$$\Delta t < 0.15 \frac{l_0}{c_0 + u_{max}}. \tag{2.21}$$

It is noted that the time step in the explicit algorithm is usually much smaller than that in the semi–implicit algorithm.

When the fluid viscosity is large the time step may also be restricted by the viscosity terms as follows [5]:

$$\Delta t < 0.25 \frac{l_0^2}{\nu}, \tag{2.22}$$

where ν is the kinematic viscosity $(\nu = \mu/\rho)$.

In this chapter the fundamentals of the MPS method are explained. Specifically the weight function, PND, discretization models, time-marching algorithms, and time-step restrictions are described. The original MPS may suffer from pressure/velocity fluctuations, instability due to negative pressure, inaccuracy of the discretization models, numerical diffusion, and so on. Further improvements will be presented in the following chapters to address these issues.

References

[1] S. Koshizuka, Y. Oka, Moving-particle semi-implicit method for fragmentation of incompressible fluid, Nucl. Sci. Eng. 123 (1996) 421−434.

[2] H. Isshiki, Discrete differential operators on irregular nodes (DDIN), Int. J. Numer. Methods Eng. 88 (2011) 1323−1343.

[3] M. Kondo, S. Koshizuka, Improvement of stability in moving particle semi-implicit method, Int. J. Numer. Methods Fluids 65 (2011) 638−654.

[4] E. Toyota, A particle method with variable spatial resolution for incompressible flows, Proc. 19th Jpn. Soc. Fluid Mech. 9 (2005) 5−10.

[5] G. Duan, B. Chen, Stability and accuracy analysis for viscous flow simulation by the moving particle semi-implicit method, Fluid Dyn. Res. 45 (2013) 1−15.

[6] K.C. Ng, Y.H. Hwang, T.W.H. Sheu, On the accuracy assessment of Laplacian models in MPS, Comput. Phys. Commun. 185 (2014) 2412−2426.

[7] G. Duan, B. Chen, Large Eddy Simulation by particle method coupled with sub-particle-scale model and application to mixing layer flow, Appl. Math. Model. 39 (2015) 3135−3149.

[8] A. Shakibaeinia, Y. Jin, A weakly compressible MPS method for modeling of open-boundary free-surface flow, Int. J. Numer. Methods Fluids 63 (2010) 1208−1232.

[9] Z. Wang, G. Duan, S. Koshizuka, A. Yamaji, Moving particle semi-implicit method, *Nuclear Power Plant Design and Analysis Codes*, Woodhead Publishing, 2001, pp. 439−461.

CHAPTER THREE

Improved discretization models

The original MPS models only have the strict convergence on the regular particle distributions. When the particle distributions become chaotic the consistency of discretization models cannot be maintained. For example, first-order derivatives (e.g., gradient and divergence) are zero order of convergence and second-order derivatives (e.g., Laplacian) are minus first order of convergence. In this chapter, several discretization schemes are introduced to improve the consistency and convergence of particle interaction models.

3.1 Improvement of pressure poisson equations

3.1.1 Compressible PPE

The source term of the original PPE is derived for incompressible fluids by assuring the particle number density (PND) n^{k+1} equal to n^0. Ikeda et al. [1] derived a compressible PPE for gases by assuming that the density has a linear relationship with pressure:

$$\frac{n_i^{n+1}}{n^0} = \frac{\rho_i^{n+1}}{\rho^0} 1 + \frac{P_i^{n+1}}{\rho^0 c^2} \qquad (3.1)$$

where c is the sound velocity. Then the PPE for the compressible fluid is written as

$$\langle \nabla^2 P \rangle_i^{n+1} = \frac{1}{\Delta t^2} \left\{ \frac{\rho(\langle n \rangle_i^* - n^0)}{n^0} - \frac{P_i^{n+1}}{c^2} \right\} \qquad (3.2)$$

The source term of the PPE represents the deviation of the particle number density n^* from $n^0 \left\{ 1 + P_i^{k+1} / (\rho^0 c^2) \right\}$.

Eq. (3.2) can also be used to the incompressible fluid by allowing a restricted compressibility to improve the numerical stability, which is reformed as follows:

$$\nabla^2 P^{k+1} = \frac{\rho}{\Delta t^2} \left(\frac{n^0 - n^*}{n^0} + \alpha P_i^{k+1} \right) \qquad (3.3)$$

Moving Particle Semi-implicit Method.
DOI: https://doi.org/10.1016/B978-0-443-13508-8.00003-2

where α is the compressibility ratio, which is usually set at $10^{-9} - 10^{-6}$. When a larger α is chosen the calculations are more stable, but the effect of compressibility is stronger and the fluid motion may be unnatural. In the calculation algorithm the second term of the source term is moved to the left hand side, and the diagonal elements of the coefficient matrix are enlarged.

On the other hand, Khayyer and Gotoh [2] deduced another weakly compressible PPE by relating pressure and density with the equation of state.

$$\left(\nabla^2 P^{k+1}\right)_i = \frac{1}{\Delta t^2 c^2}(p_{k+1} - p_k) + \frac{\rho}{\Delta t}\left(\nabla \cdot \mathbf{u}_k^*\right) \qquad (3.4)$$

where the specific value of sound velocity c is artificially determined. The value of c should be much larger than the actual speed of sound to minimize the compressibility.

3.1.2 Error compensating parts in the source term

For the incompressible fluid, two generic formulations of PPE can be derived by using different incompressible conditions: the PND condition and the divergence-free condition. When using the PND condition as the incompressible condition, that is, $\frac{D\rho}{Dt} = 0$, the PPE is written as the original equation [3]:

$$\nabla^2 P = \frac{\rho}{\Delta t^2}\frac{n^0 - n^*}{n^0} \qquad (3.5)$$

When using the divergence-free condition as the incompressible condition, that is, $\nabla \cdot \mathbf{u} = 0$, the PPE is written as

$$\nabla^2 P = \frac{\rho}{\Delta t^2}\nabla \cdot \mathbf{u}^* \qquad (3.6)$$

The PND-type PPE sustains the incompressibility by solving the difference between the present PND and the standard PND. This approach can avoid the error accumulation of PND, but the pressure oscillation is a prominent problem. In comparison the divergence-free-type PPE has the advantage of smoother pressure calculation because even though the particle distribution is nonuniform, the divergence of velocity field can be obtained smoothly. However, one problem is that the PND error may accumulate and as a result the fluid volume cannot be conserved after a long-time calculation. Therefore in order to calculate pressure smoothly and alleviate error accumulation of PND a hybrid form of PPE source term is proposed:

$$\nabla^2 P = \frac{\rho}{\Delta t^2} \nabla \cdot \mathbf{u}^* + \frac{\rho}{\Delta t^2} \frac{n^0 - n^k}{n^0} \tag{3.7}$$

However, if Eq. (3.7) is directly used to the pressure calculation the simulation is difficult to converge because the source term is too large. In order to prevent numerical divergence and sustain PND conservation the second term of the source term is multiplied by a relaxation coefficient [4].

$$\nabla^2 P = \frac{\rho}{\Delta t^2} \nabla \cdot \mathbf{u}^* + \gamma \frac{\rho}{\Delta t^2} \frac{n^0 - n^k}{n^0} \tag{3.8}$$

where γ is a relaxation coefficient which is about 1.0×10^{-3}. If γ is chosen larger, the pressure oscillation can be effectively suppressed, but the numerical convergence needs a longer time. The second term of the PPE source term in Eq. (3.8) is considered as the error compensating part for PND.

In the practical application, Eq. (3.8) is usually used together with a weakly compressible scheme to derive stable pressure calculation. Combining Eq. (3.8) and Eq. (3.3) the PPE is written as

$$\langle \nabla^2 P \rangle_i^{k+1} = (1-\gamma) \frac{\rho}{\Delta t} \nabla \cdot \mathbf{u}^* - \gamma \frac{\rho}{\Delta t^2} \left(\frac{n^* - n^0}{n^0} \right) + \alpha \frac{\rho}{\Delta t^2} P_i^{k+1} \tag{3.9}$$

On the other hand, Kondo and Koshizuka [5] deduced a generic expression of the source term directly from the hydrodynamic equations. Considering the fractional step algorithm in solving the Navier−Stokes equation the following PPE is derived:

$$-\frac{1}{\rho^0} \nabla^2 P = \frac{1}{\rho^0} \left(\frac{D^2 \rho^*}{Dt^2} \right) \tag{3.10}$$

where ρ^0 and ρ^* represent the fluid densities at the initial state and the intermediate temporary state, respectively.

On the basis that the PND is proportional to the fluid density the right-hand side of Eq. (3.10) can be approximated and discretized by the second-order differential as

$$\frac{1}{\rho^0} \left(\frac{D^2 \rho^*}{Dt^2} \right) = \frac{1}{n^0} \left(\frac{D^2 n^*}{Dt^2} \right) = \frac{1}{n^0} \frac{n^* - 2n^k + n^{k-1}}{\Delta t^2} \tag{3.11}$$

In the discretization process of Eq. (3.11) the truncation errors of PND

$$\left(e_{\left(\frac{Dn}{Dt}\right)}\right)_i^k = \frac{n_i^k - n_i^{k-1}}{\Delta t} \tag{3.12}$$

and

$$(e_{n^0})_i^k = n_i^k - n^0 \tag{3.13}$$

accumulate in a long-time calculation. To stop the error accumulation of PND, two error compensating parts are added to the source term of Eq. (3.11) [5].

$$\frac{1}{\rho^0}\left(\frac{D^2\rho}{Dt^2}\right)^{\text{other}} = \frac{1}{\rho^0}\left(\frac{D^2\rho}{Dt^2}\right)^{\text{other}} + \frac{B}{\rho^0}\left(\frac{D\rho}{Dt}\right) + \frac{\Gamma}{\rho^0}\left(\rho - \rho^0\right) \tag{3.14}$$

where B and Γ are the coefficients of the additional parts. The right-hand side of Eq. (3.14) is approximated by

$$\frac{1}{\rho^0}\left(\frac{D^2\rho^*}{Dt^2}\right) = \frac{1}{n^0}\frac{n^* - 2n^k + n^{k-1}}{\Delta t^2} + \frac{\beta}{\Delta t}\frac{n^k - n^{k-1}}{n^0 \Delta t} + \frac{\gamma}{\Delta t^2}\frac{n^k - n^0}{n^0} \tag{3.15}$$

Substituting Eq. (3.15) into Eq. (3.10) the final formation of PPE is derived as

$$-\frac{1}{\rho^0}\nabla^2 P = \frac{1}{n^0}\frac{n^* - 2n^k + n^{k-1}}{\Delta t^2} + \frac{\beta}{\Delta t}\frac{n^k - n^{k-1}}{n^0 \Delta t} + \frac{\gamma}{\Delta t^2}\frac{n^k - n^0}{n^0} \tag{3.16}$$

where β and γ are coefficients to adjust compensating parts, and the relationship of them should be kept as

$$0 \le \gamma \le \beta \le 1. \tag{3.17}$$

However, the determination of β and γ is a cumbersome procedure because they are affected by particle space distance and time step. If the combination of β and γ is not chosen appropriately, the pressure may oscillate seriously. Kondo and Koshizuka [5] suggest a pair of empirical values after a series of tests, which is $(\beta, \gamma) = (0.05, 0.0005)$ for the conditions of $l_0 = 0.012m$ and $\Delta t = 1.0 \times 10^{-4}$ s.

Khayyer and Gotoh [6] pointed out that the adjusting coefficients β and γ in the PPE source term developed by Kondo and Koshizuka [5] are calibrated from a hydrostatic pressure calculation, which may not be appropriate in the simulation of violent fluid flows. Moreover the numerical schemes applied for approximations of first-order and second-order time derivatives of PND are first-order-accurate ones, which may introduce further numerical errors in the practical simulations of violent fluid flows. Therefore a new formulation of PPE is proposed considering the dynamic coefficients for error-compensating parts.

$$
\begin{cases}
\left(\dfrac{\Delta t}{\rho}\nabla^2 p_{k+1}\right) = \dfrac{1}{n_0}\left(\dfrac{Dn}{Dt}\right)^*_i + \left|\left(\dfrac{n^k-n_0}{n_0}\right)\right|\left[\dfrac{1}{n_0}\left(\dfrac{Dn}{Dt}\right)^k_i\right] \\
\qquad + \left|\left(\dfrac{\Delta t}{n_0}\left(\dfrac{Dn}{Dt}\right)^k_i\right)\right|\left[\dfrac{1}{\Delta t}\left(\dfrac{n^k-n_0}{n_0}\right)\right], \quad n^k > \gamma n_0 \\[2em]
\left(\dfrac{\Delta t}{\rho}\nabla^2 p_{k+1}\right) = \dfrac{1}{n_0}\left(\dfrac{Dn}{Dt}\right)^*_i + \left|\left(\dfrac{n^k-(n_i)_S}{(n_i)_S}\right)\right|\left[\dfrac{1}{n_0}\left(\dfrac{Dn}{Dt}\right)^k_i\right] \\
\qquad + \left|\left(\dfrac{\Delta t}{n_0}\left(\dfrac{Dn}{Dt}\right)^k_i\right)\right|\left[\dfrac{1}{\Delta t}\left(\dfrac{n^k-(n_i)_S}{(n_i)_S}\right)\right], \quad n^k \le \gamma n_0
\end{cases}
$$

$$(3.18)$$

where different PPE source terms are applied for the internal and free surface particles. Parameter $(n_i)_S$ corresponds to the initial PND of a target particle i, which is initially located at or close to the free surface. It is updated every certain time steps to consider particle motion in the violent fluid flow. The absolute values of coefficients denote the intensity of error-compensating term variations, whereas the negative or positive signs are determined by the error-compensating terms themselves. Compared with Eq. (3.16) using the fixed coefficients for error-compensating parts, Eq. (3.18) considers both the variation speed of PND and its deviation amplitude from the standard value.

3.1.3 High-order PPE

Khayyer and Gotoh [2] revisited PPE derivation process, and the standard PPE is revised as

$$
\langle \nabla^2 P \rangle^{k+1}_i = \frac{\rho}{\Delta t^2}\frac{n^0-n^*}{n^0} = \frac{\rho}{\Delta t \cdot n^0}\frac{Dn}{Dt}
\tag{3.19}
$$

The time differential of PND is calculated using the original MPS weight function.

$$\frac{Dn}{Dt} = \sum_{i \neq j} \frac{Dw(|r_j - r_i|)}{Dt}$$ (3.20)

Considering the original MPS weight function (Eq. 2.3) a two-dimensional (2D) estimation of the PND differential is

$$\frac{Dw(|r_j - r_i|)}{Dt} = \frac{Dw_{ij}}{Dt} = \left(\frac{\partial w_{ij}}{\partial r_{ij}} \frac{\partial r_{ij}}{\partial x_{ij}} \frac{dx_{ij}}{dt} + \frac{\partial w_{ij}}{\partial r_{ij}} \frac{\partial r_{ij}}{\partial y_{ij}} \frac{dy_{ij}}{dt} \right)$$

$$= \left(\frac{-r_e}{r_{ij}^2} \frac{x_{ij}}{r_{ij}} u_{ij}^* + \frac{-r_e}{r_{ij}^2} \frac{y_{ij}}{r_{ij}} v_{ij}^* \right) = \frac{-r_e}{r_{ij}^3} \left(x_{ij} u_{ij}^* + y_{ij} v_{ij}^* \right)$$ (3.21)

Therefore the PPE with the high-order source term is written as

$$\left\langle \nabla^2 P_{k+1} \right\rangle_i = - \frac{\rho}{n_0 \Delta t} \left(\sum_{j \neq i} \frac{r_e}{r_{ij}^3} (x_{ij} u_{ij} + y_{ij} v_{ij}) \right)^*$$ (3.22)

In three dimensions the high-order source term is expressed as

$$\left(\nabla^2 P_{k+1} \right)_i = \frac{\rho}{n_0 \Delta t} \left(\frac{Dn}{Dt} \right)^* = - \frac{\rho}{n_0 \Delta t} \left(\sum_{i \neq j} \frac{r_e}{r_{ij}^3} \left(x_{ij} u_{ij} + y_{ij} v_{ij} + z_{ij} w_{ij} \right)^* \right)$$ (3.23)

In addition a high-order formulation of Laplacian model is also developed for the discretization of PPE [7,8]. The Laplacian is defined as the divergence of the gradient at a target particle i.

$$\left\langle \nabla^2 \phi \right\rangle_i = \nabla \cdot \left\langle \nabla \phi \right\rangle_i$$ (3.24)

The gradient at a target particle i can be written in the following form:

$$\left\langle \nabla^2 \phi \right\rangle_i = \frac{1}{\sum_{i \neq j} w_{ij}} \sum_{i \neq j} \left(\phi_j - \phi_i \right) \nabla w_{ij} = \frac{1}{\sum_{i \neq j} w_{ij}} \sum_{i \neq j} \phi_{ij} \nabla w_{ij}$$ (3.25)

where ϕ is an arbitrary scalar variable, and $\phi_{ij} = \phi_j - \phi_i$. Substituting Eq. (3.25) into Eq. (3.24) the following equation can be obtained:

$$\nabla \cdot \left\langle \nabla \phi \right\rangle_i = \frac{1}{n_0} \sum_{i \neq j} \left(\nabla \phi_{ij} \cdot \nabla w_{ij} + \phi_{ij} \nabla^2 w_{ij} \right)$$ (3.26)

where the gradients of ϕ_{ij} and w_{ij} are expressed as follows in the 2D Cartesian coordinates:

$$\nabla \phi_{ij} = \frac{\partial \phi_{ij}}{\partial r_{ij}} \frac{\partial r_{ij}}{\partial x_{ij}} \mathbf{i} + \frac{\partial \phi_{ij}}{\partial r_{ij}} \frac{\partial r_{ij}}{\partial y_{ij}} \mathbf{j} \tag{3.27}$$

$$\nabla w_{ij} = \frac{\partial w_{ij}}{\partial r_{ij}} \frac{\partial r_{ij}}{\partial x_{ij}} \mathbf{i} + \frac{\partial w_{ij}}{\partial r_{ij}} \frac{\partial r_{ij}}{\partial y_{ij}} \mathbf{j} \tag{3.28}$$

and

$$\nabla \phi_{ij} \cdot \nabla w_{ij} = \frac{\partial \phi_{ij}}{\partial r_{ij}} \frac{\partial w_{ij}}{\partial r_{ij}} = \frac{\phi_{ji} - \phi_{ij}}{r_{ij}} \frac{\partial w_{ij}}{\partial r_{ij}} = \frac{2\phi_{ji}}{r_{ij}} \frac{\partial w_{ij}}{\partial r_{ij}} \tag{3.29}$$

On the other hand,

$$\nabla^2 w_{ij} = \nabla \cdot \nabla w_{ij} = \frac{\partial^2 w_{ij}}{\partial r_{ij}^2} \left(\frac{x_{ij}^2}{r_{ij}^2} + \frac{y_{ij}^2}{r_{ij}^2} \right) + \left(\frac{2}{r_{ij}} \frac{\partial w_{ij}}{\partial r_{ij}} - \frac{1}{r_{ij}} \frac{\partial w_{ij}}{\partial r_{ij}} \right) \tag{3.30}$$

Substituting Eqs. (3.29) and (3.30) into Eq. (3.26) the following Laplacian formulation is derived

$$\langle \nabla^2 \phi \rangle_i = \nabla \cdot \langle \nabla \phi \rangle_i = \frac{1}{n_0} \sum_{j \neq i} \left(\frac{2\phi_{ji}}{r_{ij}} \frac{\partial w_{ij}}{\partial r_{ij}} + \phi_{ji} \frac{\partial^2 w_{ij}}{\partial r_{ij}^2} + \frac{\phi_{ij}}{r_{ij}} \frac{\partial w_{ij}}{\partial r_{ij}} \right)$$
$$= \frac{1}{n_0} \sum_{j \neq i} \left(\phi_{ji} \frac{\partial^2 w_{ij}}{\partial r_{ij}^2} - \frac{\phi_{ji}}{r_{ij}} \frac{\partial w_{ij}}{\partial r_{ij}} \right) \tag{3.31}$$

Using the original MPS weight function (Eq. 2.3) the high-order Laplacian model in two dimensions is written as

$$\nabla \cdot \langle \nabla \phi \rangle_i = \frac{1}{n_0} \sum_{i \neq j} \left(\frac{3\phi_{ij} r_e}{r_{ij}^3} \right) \tag{3.32}$$

When in three dimensions the high-order Laplacian model is expressed as

$$\nabla \cdot \langle \nabla \phi \rangle_i = \frac{1}{n_0} \sum_{i \neq j} \left(\frac{2\phi_{ij} r_e}{r_{ij}^3} \right) \tag{3.33}$$

Considering the convenient application the above 2D and 3D high-order PPEs can be written to a generic form.

$$\frac{1}{n_0}\sum_{j\neq i}\frac{6(P_j^{k+1}-P_i^{k+1})r_e}{d\left|\mathbf{r}_j^*-\mathbf{r}_i^*\right|^3} = -\frac{\rho}{n_0\Delta t}\sum_{j\neq i}\frac{r_e}{\left|\mathbf{r}_j^*-\mathbf{r}_i^*\right|^3}(\mathbf{r}_j^*-\mathbf{r}_i^*)(\mathbf{u}_j^*-\mathbf{u}_i^*)$$

(3.34)

It should be noted that the above high-order formulations are deduced based on the original weight function. When one uses a different weight function for the particle interaction models the formulations should be changed accordingly.

3.2 Corrective matrix for particle interaction models

3.2.1 First-order corrective matrix

Randles and Libersky [9] first proposed the corrective matrix for the smooth particle hydrodynamics models. In the MPS method, Suzuki [10] and Khayyer and Gotoh [6] derived the corrective matrix to restore the consistency of the gradient model.

In this section the discretization models using the corrective matrix developed by Duan et al. [11] are presented. These formulations take the following advantages: First the corrective matrix is dimensionless and independent from the particle size. Second, all the gradient, divergence, and Laplacian models rather than only the gradient one are corrected by the corrective matrix. For example the corrected gradient and divergence models exhibit the first order of convergence. The corrected Laplacian model takes the zero order of convergence, while the original model has the minus first order. Third the corrected models are still quite similar to the original models, and only slight modifications are necessary. In the corrected models the capability of the corrective matrix is fully exploited.

Even though there are different methods to derive the corrective matrix [6,9,10], a general way is to use the least squares (LS) method. For example the first-order Taylor series expansion between two particles in 2D is as follows:

$$\phi_j \approx \phi_i + \phi_x x_{ij} + \phi_y y_{ij}$$

(3.35)

where ϕ_x is the partial derivative with respect to x, $x_{ij} = x_j - x_i$, and $y_{ij} = y_j - y_i$. The particle distance can be calculated from $r_{ij} = \sqrt{x_{ij}^2 + y_{ij}^2}$. Then the expansion can be rewritten as

$$\frac{\left(\phi_j - \phi_i\right)}{r_{ij}} \approx \phi_x \frac{x_{ij}}{r_{ij}} + \phi_y \frac{y_{ij}}{r_{ij}} \tag{3.36}$$

where $(\phi_j - \phi_i)/r_{ij}$ is the directional derivative between particle i and j. Eq. (3.36) can be selected as the fitting function in the LS methods. For example the unknown partial derivatives can be obtained by minimizing the error between the observed and estimated directional derivatives. In this manner the corrective matrix can be obtained. When the first order Taylor series expansion is employed the first-order corrective matrix (FCM) is derived. The derivation can be found in [12].

The FCM can be calculated as follows:

$$\mathbf{C}_i^{-1} = \frac{d}{n^0} \begin{bmatrix} \sum_{j \neq i} \frac{x_{ij}^2}{r_{ij}^2} w_{ij} & \sum_{j \neq i} \frac{x_{ij} y_{ij}}{r_{ij}^2} w_{ij} \\ \sum_{j \neq i} \frac{y_{ij} x_{ij}}{r_{ij}^2} w_{ij} & \sum_{j \neq i} \frac{y_{ij}^2}{r_{ij}^2} w_{ij} \end{bmatrix} \tag{3.37}$$

where $\mathbf{C}_i$ is the corrective matrix and w_{ij} is the weight function between particle i and j, namely, $w_{ij} = w(r_{ij}, r_e)$. It is noted that $\mathbf{C}_i$ is dimensionless.

Based on the corrective matrix the corrected gradient model becomes

$$\langle \nabla \phi \rangle_i = \frac{d}{n^0} \sum_{j \neq i} \left\{ \frac{\left(\phi_j - \phi_i\right)}{r_{ij}} \left(\mathbf{C}_i \frac{\mathbf{r}_j - \mathbf{r}_i}{r_{ij}} \right) w_{ij} \right\} \tag{3.38}$$

where $\mathbf{r}_i$ is the position vector of particle i, namely, $\mathbf{r}_i = (x_i, y_i)$. In Eq. (3.38) the corrective matrix is only used to modify the unit direction vector between particle i and j. For convenience the position vector is written as a column vector for the matrix-vector multiplication in Eq. (3.38).

Divergence is also composed of first-order partial derivatives, which has been computed in the gradient modes. Based on some mathematical manipulations the corrected divergence model can be similarly written as

$$\langle \nabla \cdot \mathbf{u} \rangle_i = \frac{d}{n^0} \sum_{j \neq i} \left\{ \frac{(\mathbf{u}_j - \mathbf{u}_i)}{r_{ij}} \left(\mathbf{C}_i \frac{\mathbf{r}_j - \mathbf{r}_i}{r_{ij}} \right) w_{ij} \right\} \tag{3.39}$$

where $\mathbf{u}$ is the velocity vector. In this situation the corrective matrix is also only used to correct the unit direction vector. For example, after the corrected direction vector is employed the consistency can be restored.

However, the FCM cannot be directly employed to the Laplacian model because Laplacian is composed of second-order derivatives. Nevertheless, FCM can help to eliminate the error caused by first-order derivatives in the Laplacian model. In this situation the second-order Taylor series expansion is used:

$$\phi_j - \phi_i = \phi_x x_{ij} + \phi_y y_{ij} + \frac{1}{2}\phi_{xx} x_{ij}^2 + \frac{1}{2}\phi_{xy} x_{ij} y_{ij} + \frac{1}{2}\phi_{yy} y_{ij}^2 \qquad (3.40)$$

where ϕ_{xx} is the second-order partial derivative with respect to x and so on. Taking Eq. (3.40) into the original Laplacian model in Chapter 2, one can have the following relation:

$$\frac{2d}{n^0\lambda}\sum_{j\neq i}\left\{\left(\phi_j - \phi_i\right)w_{ij}\right\} = \phi_x \frac{2d}{n^0\lambda}\sum_{j\neq i}x_{ij}w_{ij} + \phi_y \frac{2d}{n^0\lambda}\sum_{j\neq i}y_{ij}w_{ij} + $$
$$\phi_{xx}\frac{d}{n^0\lambda}\sum_{j\neq i}x_{ij}^2 w_{ij} + \phi_{xy}\frac{2d}{n^0\lambda}\sum_{j\neq i}x_{ij}y_{ij}w_{ij} + \phi_{yy}\frac{d}{n^0\lambda}\sum_{j\neq i}y_{ij}^2 w_{ij} \qquad (3.41)$$

To effectively represent Laplacian the following equations should hold:

$$\frac{2d}{n^0\lambda}\sum_{j\neq i}x_{ij}w_{ij} \approx 0, \frac{2d}{n^0\lambda}\sum_{j\neq i}y_{ij}w_{ij} \approx 0 \qquad (3.42)$$

and

$$\frac{d}{n^0\lambda}\sum_{j\neq i}x_{ij}^2 w_{ij} \approx 1, \frac{2d}{n^0\lambda}\sum_{j\neq i}x_{ij}y_{ij}w_{ij} \approx 0, \frac{d}{n^0\lambda}\sum_{j\neq i}y_{ij}^2 w_{ij} \approx 1 \qquad (3.43)$$

On regular particle distributions, Eqs. (3.42) and (3.43) can exactly hold. In this situation the Laplacian model will take second order of convergence. However, on the chaotic particle distributions, these equations only hold approximately, causing error to the Laplacian model. The errors caused by Eq. (3.42) can be called the first-derivative errors for convenience, and they are the main error parts in the Laplacian model based on a dimension analysis [11]. The errors caused by Eq. (3.43) can be called the second-derivative errors.

Because the ϕ_x and ϕ_y have been accurately calculated in the corrected gradient model, they can be used to evaluate the first-derivative errors. Then the first-derivative errors can be removed in the total errors of the original Laplacian model. This can help to improve the accuracy of the Laplacian mode. For convenience the following (row) vector $\mathbf{L}_i$ composed of the coefficients in Eq. (3.42) is defined for each particle:

$$\mathbf{L}_i = \left(\frac{2d}{n^0 \lambda} \sum_{j \neq i} x_{ij} w_{ij}, \frac{2d}{n^0 \lambda} \sum_{j \neq i} \gamma_{ij} w_{ij} \right) \tag{3.44}$$

Note that the inner product of $\mathbf{L}_i$ and $\langle \nabla \phi \rangle_i$ is the first-derivative error. By subtracting $\mathbf{L}_i \langle \nabla \phi \rangle_i$ from both sides of Eq. (3.41), one can obtain a corrected Laplacian model:

$$\langle \nabla^2 \phi \rangle_i = \frac{2d}{n^0 \lambda} \sum_{j \neq i} \left\{ \left(\phi_j - \phi_i \right) w_{ij} \right\} - \mathbf{L}_i \cdot \langle \nabla \phi \rangle_i \tag{3.45}$$

Taking the corrected gradient model into the above equation, one can simplify the Laplacian model as follows:

$$\langle \nabla^2 \phi \rangle_i = \frac{d}{n^0} \sum_{j \neq i} \left\{ \left(\phi_j - \phi_i \right) \left(\frac{2}{\lambda} - \frac{\mathbf{L}_i \mathbf{C}_i \left(\mathbf{r}_j - \mathbf{r}_i \right)}{r_{ij}^2} \right) w_{ij} \right\} \tag{3.46}$$

It is noted that the $\mathbf{L}_i \mathbf{C}_i \left(\mathbf{r}_j - \mathbf{r}_i \right) / r_{ij}^2$ is a dimensionless scalar. For the computation convenience, $\mathbf{L}_i$, $\mathbf{C}_i$, and $\left(\mathbf{r}_j - \mathbf{r}_i \right)$ are a row vector, a matrix, and a column vector, respectively. The errors of the FCM schemes are analyzed in Ref. [12].

The convergence behaviors of the aforementioned FCM models are presented in Fig. 3.1 from Ref. [11]. The convergence tests were performed on the chaotic particle distributions. The Brookshaw-type model is from Ref. [13]. The Khayyer—Gotoh-type model is from Ref [7]. The situation of the divergence model is exactly like that of the gradient model. The original gradient model is of the zero order of convergence. The FCM has improved the accuracy of the corrected gradient model to first order of convergence. Meanwhile the convergence orders of the original Laplacian models (including Brookshaw-type and Khayyer—Gotoh-type) are around -1. When the FCM is used the convergence order is raised to zero order. Briefly the accuracy is improved remarkably by the FCM.

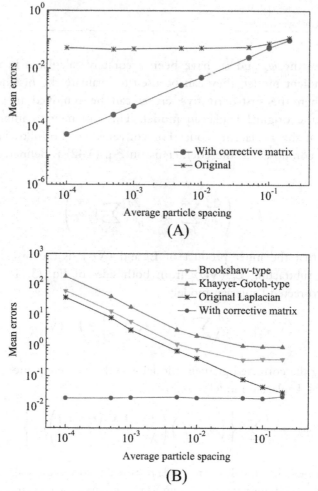

Figure 3.1 Mean errors of different discretization models with respect to average particle spacing [11]. (A) Gradient models; (B) Laplacian models.

The corrected models in 3D are like the 2D models. The 3D corrected models are presented below without derivation. First the corrective matrix in 3D is calculated from

$$
\mathbf{C}_i^{-1} = \frac{d}{n^0} \begin{bmatrix} \sum_{j \neq i} \frac{x_{ij}^2}{r_{ij}^2} w_{ij} & \sum_{j \neq i} \frac{x_{ij}y_{ij}}{r_{ij}^2} w_{ij} & \sum_{j \neq i} \frac{x_{ij}z_{ij}}{r_{ij}^2} w_{ij} \\ \sum_{j \neq i} \frac{x_{ij}y_{ij}}{r_{ij}^2} w_{ij} & \sum_{j \neq i} \frac{y_{ij}^2}{r_{ij}^2} w_{ij} & \sum_{j \neq i} \frac{y_{ij}z_{ij}}{r_{ij}^2} w_{ij} \\ \sum_{j \neq i} \frac{x_{ij}z_{ij}}{r_{ij}^2} w_{ij} & \sum_{j \neq i} \frac{y_{ij}z_{ij}}{r_{ij}^2} w_{ij} & \sum_{j \neq i} \frac{z_{ij}^2}{r_{ij}^2} w_{ij} \end{bmatrix} \qquad (3.47)
$$

Meanwhile vector $\mathbf{L}_i$ for the Laplacian model is calculated as follows:

$$\mathbf{L}_i = \left(\frac{2d}{n^0 \lambda} \sum_{j \neq i} x_{ij} w_{ij}, \; \frac{2d}{n^0 \lambda} \sum_{j \neq i} y_{ij} w_{ij}, \; \frac{2d}{n^0 \lambda} \sum_{j \neq i} z_{ij} w_{ij} \right) \qquad (3.48)$$

With the above definition, the gradient, divergence, and Laplacian models in 3D can be computed from Eqs. (3.38), (3.39), and (3.46), respectively.

3.2.2 Second-order corrective matrix

The second-order corrective matrix (SCM) can be derived from the LS method, as presented by Duan et al. [12]. During the LS process a key point is to select the fitting function. Like the fitting function of FCM in Eq. (3.36), the fitting function of SCM is as follows:

$$\begin{aligned} \frac{\left(\phi_j - \phi_i \right)}{r_{ij}} &\approx \phi_x \frac{x_{ij}}{r_{ij}} + \phi_y \frac{y_{ij}}{r_{ij}} + \frac{l_0}{2} \phi_{xx} \frac{x_{ij}^2}{l_0 r_{ij}} \\ &+ \frac{l_0}{2} \phi_{yy} \frac{y_{ij}^2}{l_0 r_{ij}} + l_0 \phi_{xy} \frac{x_{ij} y_{ij}}{l_0 r_{ij}} \end{aligned} \qquad (3.49)$$

Taking the dimensionless w_{ij}/n^0 as the weight for each neighboring particle j the LS method can be used to estimate all the first and second partial derivatives, from which the gradient, divergence, and Laplacian models can be constituted. The detailed derivation process can be found in Ref. [12]. There are two main advantages of the fitting function in Eq. (3.49). First the derived high-order schemes are still quite like the original MPS models [12]. The main modifications come from the corrected direction vectors, as discussed later. Second the Neumann boundary conditions can be easily coupled with this fitting function because the dimension of Neumann boundary condition is exactly the same with the dimension of Eq. (3.49). The coupling of the Neumann boundary condition into the SCM can be found in Ref. [14].

The SCM in 2D is calculated in the following manner: A generalized second-order direction vector $\mathbf{P}$ is defined between the particle pair i and j as follows:

$$\mathbf{P} = \left[\frac{x_{ij}}{r_{ij}}, \frac{y_{ij}}{r_{ij}}, \frac{x_{ij}^2}{l_0 r_{ij}}, \frac{y_{ij}^2}{l_0 r_{ij}}, \frac{x_{ij} y_{ij}}{l_0 r_{ij}} \right]^T = \left[p_1, p_2, p_3, p_4, p_5 \right]^T \qquad (3.50)$$

where $\mathbf{P}$ is a column vector. Then the SCM is computed as follows:

$$
\mathbf{C} = \begin{bmatrix} (P_1, P_1) & (P_1, P_2) & \cdots & (P_1, P_5) \\ (P_2, P_1) & (P_2, P_2) & \cdots & (P_2, P_5) \\ \vdots & \vdots & \ddots & \vdots \\ (P_5, P_1) & (P_5, P_2) & \cdots & (P_5, P_5) \end{bmatrix}^{-1} \tag{3.51}
$$

where (P_α, P_β) is an inner product of components P_α and P_β in vector $\mathbf{P}$ by looping all neighbor particles using the previous weight w_{ij}/n^0. The inner product is calculated as follows:

$$
(P_\alpha, P_\beta) = \sum_{j \neq i} P_\alpha \cdot P_\beta \frac{w_{ij}}{n^0} \tag{3.52}
$$

Here an example to calculate (P_2, P_3) is shown below:

$$
(P_2, P_3) = \sum_{j \neq i} P_2 \cdot P_3 \frac{w_{ij}}{n^0} = \sum_{j \neq i} \frac{y_{ij}}{r_{ij}} \frac{x_{ij}^2}{l_0 r_{ij}} \frac{w_{ij}}{n^0} \triangleleft \tag{3.53}
$$

It must be noted that both the generalized direction vector $\mathbf{P}$ and the corrective matrix $\mathbf{C}$ are dimensionless. This helps to suppress the large condition number that may cause difficulty in calculating matrix inversion [15]. The product of $\mathbf{C}$ and $\mathbf{P}$ is called the dimensionless *"generalized corrected direction vector,"* which makes the high–order discretization models still like the original models.

As derived in Ref. [12] the high-order gradient model based on SCM is as follows:

$$
\langle \nabla \phi \rangle_i = \frac{1}{n^0} \sum_{j \neq i} \left\{ w_{ij} \frac{\phi_j - \phi_i}{r_{ij}} \left(\begin{bmatrix} \mathbf{C}_1 \\ \mathbf{C}_2 \end{bmatrix} \mathbf{P} \right) \right\} \tag{3.54}
$$

where $\mathbf{C}_1$ and $\mathbf{C}_2$ are the first and second rows of matrix $\mathbf{C}$, respectively. Compared to the original gradient model or the FCM gradient model (Eq. 3.38) the product of $[\mathbf{C}_1 \cdot \mathbf{C}_2]^T$ and $\mathbf{P}$ is actually the *real corrected direction vector*, which just corresponds to the first two components of the *generalized corrected direction vector* (namely, product of $\mathbf{C}$ and $\mathbf{P}$). In this situation the influence of chaotic particle distributions is just considered in the corrected direction vector. In this manner the accuracy is greatly enhanced. Similarly the high-order divergence model based on SCM is as follows:

$$
\langle \nabla \cdot \mathbf{u} \rangle_i = \frac{1}{n^0} \sum_{j \neq i} \left\{ w_{ij} \frac{\mathbf{u}_j - \mathbf{u}_i}{r_{ij}} \left(\begin{bmatrix} \mathbf{C}_1 \\ \mathbf{C}_2 \end{bmatrix} \mathbf{P} \right) \right\} \tag{3.55}
$$

It is noted that the same corrected direction vector is used in the high-order gradient and divergence models.

Based on the SCM, it is possible to derive the strictly convergent Laplacian model. The third and fourth components of the *generalized corrected direction vector* just correspond to the second partial derivatives. The high-order Laplacian model based on SCM is as follows:

$$\langle \nabla^2 \phi \rangle_i = \frac{2}{n^0} \sum_{j \neq i} \left\{ w_{ij} \left(\phi_j - \phi_i \right) \frac{[C_3 + C_4]P}{l_0 r_{ij}} \right\} \tag{3.56}$$

where C_3 and C_4 are the third and fourth rows of matrix C. Note that the product of $[C_3 + C_4]$ and P is a dimensionless scalar. The effects of anisotropy of particle distributions are considered in this dimensionless scalar. The errors of the SCM schemes are analyzed in Ref. [12].

The error comparison among the original, FCM, and SCM discretization models is investigated numerically in Ref. [12]. The quasirandom particle distribution is generated as follows: First a uniform Cartesian particle distribution with constant particle spacing, l_0, is produced. Second, some random displacements up to $0.15 \times l_0$ are added. Each test case is calculated for 500 times, and the mean errors are compared. The original models are from Section 2.2. The FCM models are from Section 3.2.1. The error comparison of different models is depicted in Fig. 3.2. When the particle spacing is large, both gradient and divergence models exhibit the second-order convergence because the smoothing error is dominant. However, when the particle spacing becomes small the observed convergence order is consistent with the theoretical analysis. Specifically, first the original, FCM, and SCM gradient/divergence models exhibit zero, first, and second order of convergence, respectively. Second the original, FCM, and SCM Laplacian models exhibit minus first, zero, and first order of convergence, respectively. It is obvious that the SCM models significantly reduce the discretization error compared to the original or FCM models.

Next the SCM schemes in 3D are presented below. The schemes in 3D are still like the ones in 2D above. The generalized column vector P between particles i and j is defined as follows:

$$P = \left[\frac{x_{ij}}{r_{ij}}, \frac{y_{ij}}{r_{ij}}, \frac{z_{ij}}{r_{ij}}, \frac{x_{ij}^2}{l_0 r_{ij}}, \frac{y_{ij}^2}{l_0 r_{ij}}, \frac{z_{ij}^2}{l_0 r_{ij}}, \frac{x_{ij}y_{ij}}{l_0 r_{ij}}, \frac{x_{ij}z_{ij}}{l_0 r_{ij}}, \frac{y_{ij}z_{ij}}{l_0 r_{ij}} \right]^T \tag{3.57}$$

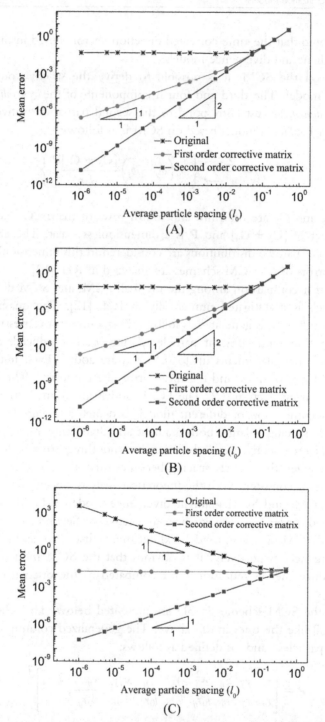

Figure 3.2 Mean errors of different discretization models with respect to average particle spacing (l_0) [12]. (A) Gradient models. (B) Divergence models. (C) Laplacian models.

where $\mathbf{P}$ is a column vector. This vector can also be written as

$$\mathbf{P} = \left[p_1, p_2, p_3, p_4, p_5, p_6, p_7, p_8, p_9 \right]^T \qquad (3.58)$$

for the convenience of the SCM calculation. Then the SCM in 3D can be computed from

$$\mathbf{C} = \begin{bmatrix} (P_1, P_1) & (P_1, P_2) & \cdots & (P_1, P_9) \\ (P_2, P_1) & (P_2, P_2) & \cdots & (P_2, P_9) \\ \vdots & \vdots & \ddots & \vdots \\ (P_9, P_1) & (P_9, P_2) & \cdots & (P_9, P_9) \end{bmatrix}^{-1} \qquad (3.59)$$

where the definition of the inner product $\left(P_\alpha, P_\beta \right)$ is the same with Eq. (3.52) in 2D. The influence of the chaotic particle distributions is also considered in the product of $\mathbf{C}$ and $\mathbf{P}$, similar to 2D.

The high-order gradient model based on SCM in 3D can be calculated as follows:

$$\langle \nabla \phi \rangle_i = \frac{1}{n^0} \sum_{j \neq i} \left\{ w_{ij} \frac{\phi_j - \phi_i}{r_{ij}} \left(\begin{bmatrix} \mathbf{C}_1 \\ \mathbf{C}_2 \\ \mathbf{C}_3 \end{bmatrix} \mathbf{P} \right) \right\} \qquad (3.60)$$

where $\mathbf{C}_1$, $\mathbf{C}_2$, and $\mathbf{C}_3$ are the first, second, and third rows of matrix $\mathbf{C}$, respectively. The product of $[\mathbf{C}_1 \cdot \mathbf{C}_2 \cdot \mathbf{C}_3]^T$ and $\mathbf{P}$ is just the corrected direction vector. Similarly the high-order divergence model in 3D is derived as follows:

$$\langle \nabla \cdot \mathbf{u} \rangle_i = \frac{1}{n^0} \sum_{j \neq i} \left\{ w_{ij} \frac{\mathbf{u}_j - \mathbf{u}_i}{r_{ij}} \left(\begin{bmatrix} \mathbf{C}_1 \\ \mathbf{C}_2 \\ \mathbf{C}_3 \end{bmatrix} \mathbf{P} \right) \right\} \qquad (3.61)$$

The fourth to sixth components of the product of $\mathbf{C}$ and $\mathbf{P}$ correspond to the second-order partial derivatives of the Laplacian. Based on these components the high-order Laplacian model in 3D can be written as

$$\langle \nabla^2 \phi \rangle_i = \frac{2}{n^0} \sum_{j \neq i} \left\{ w_{ij} \left(\phi_j - \phi_i \right) \frac{[\mathbf{C}_4 + \mathbf{C}_5 + \mathbf{C}_6] \mathbf{P}}{l_0 r_{ij}} \right\} \qquad (3.62)$$

where $\mathbf{C}_4$ to $\mathbf{C}_6$ are the fourth to sixth rows of matrix $\mathbf{C}$, respectively. Again the product of $[\mathbf{C}_4 + \mathbf{C}_5 + \mathbf{C}_6]$ and $\mathbf{P}$ in Eq. (3.62) is a dimensionless scalar.

It must be noted that the use of the high-order models near free surface severely depends on the free-surface detection method. Specifically, when the conventional free-surface detection method is employed the FCM and SCM discretization models can easily trigger instability. In this situation, it is better to apply the FCM/SCM discretization models for the internal particles and apply the original discretization models near and at free surfaces [11]. When the advanced free-surface detection method based on the model coefficients [16] is employed the high-order schemes can be applied near and at free surfaces. Meanwhile the boundary conditions must be imposed precisely [16].

3.3 Least square MPS

In this section the least square MPS (LSMPS), an arbitrary high-order mesh-free spatial discretization scheme developed by Tamai and Koshizuka [17], is introduced. In general the LSMPS scheme contains two formulations: type A and type B. Type A scheme is used to evaluate the differential operators on a particle. In contrast the type B scheme calculates differential operators at an arbitrary position.

In type A scheme a Taylor series expansion of a scalar variable ϕ around $\mathbf{r}_i$ with a nearby particle $\mathbf{r}_j$ is written as

$$\phi_j = \phi_i + \sum_{m=1}^{p} \frac{1}{m!} \left(\mathbf{r}_{ij} \cdot \nabla \right)^m \phi|_{\mathbf{r}_i} + \mathrm{o}\left(\left| \mathbf{r}_{ij} \right|^{p+1} \right) \tag{3.63}$$

If the higher-order terms are ignored, the following equation is obtained:

$$\phi_j - \phi_i = \sum_{m=1}^{p} \frac{r_s^m}{m!} \left(\frac{\mathbf{r}_{ij}}{r_s} \cdot \nabla \right)^m \phi|_{\mathbf{r}_i} = \mathbf{p}_{ij} \cdot \left(\mathbf{H}_{r_s}^{-1} \partial \phi|_{\mathbf{r}_i} \right) \tag{3.64}$$

where $\phi_i = \phi(\mathbf{r}_i)$, $\phi_j = \phi(\mathbf{r}_j)$, $\mathbf{r}_{ij} = \mathbf{r}_j - \mathbf{r}_i$, r_s denotes a scaling parameter, p is a positive integer, which indicates the order of approximation, $\mathbf{p}_{ij}$ denotes a polynomial basis vector, $\mathbf{H}_{r_s}^{-1}$ represents a scaling matrix, and ∂ indicates a differential operator vector.

For the second-order scheme ($p = 2$) in two dimensions, these functions are written as

$$\mathbf{p}_{ij} = \mathbf{p}\left(\frac{\mathbf{r}_{ij}}{r_s} \right) = \left[\frac{x_{ij}}{r_s}, \frac{y_{ij}}{r_s}, \frac{x_{ij}^2}{r_s^2}, \frac{y_{ij}^2}{r_s^2}, \frac{x_{ij}y_{ij}}{r_s^2} \right]^T \tag{3.65}$$

$$\mathbf{H}_{r_s} = \text{diag}\left(\frac{1}{r_s}, \frac{1}{r_s}, \frac{2}{r_s^2}, \frac{2}{r_s^2}, \frac{1}{r_s^2}\right) \tag{3.66}$$

$$\partial = \left[\frac{\partial}{\partial x}, \frac{\partial}{\partial y}, \frac{\partial^2}{\partial x^2}, \frac{\partial^2}{\partial y^2}, \frac{\partial^2}{\partial x \partial y}\right]^T \tag{3.67}$$

By considering all neighboring particles and nodes as j ($j \neq i$) for Eq. (3.64) the objective function J for a weighted least squares problem is defined as

$$J(\mathbf{X}_i) = \sum_{j \neq i} w_{ij}\left(\mathbf{p}_{ij} \cdot \mathbf{X}_i - \phi_j + \phi_i\right)^2 \tag{3.68}$$

with

$$\mathbf{X}_i = \mathbf{H}_{r_s}^{-1}\partial\phi|_{\mathbf{r}_i} \tag{3.69}$$

The following weight function w is widely adopted in the LSMPS method:

$$w\left(|\mathbf{r}_{ij}|\right) = \begin{cases} \left(1 - \frac{|\mathbf{r}_{ij}|}{r_e}\right)^2, & |\mathbf{r}_{ij}| < r_e \\ 0, & |\mathbf{r}_{ij}| \geq r_e \end{cases} \tag{3.70}$$

where r_e indicates the effective radius, which is usually set to 3.1 l_0 for the second-order scheme.

Minimization of J leads to the normal equations:

$$\mathbf{M}_i\mathbf{X}_i = \mathbf{b}_i \tag{3.71}$$

with

$$\mathbf{M}_i = \sum_{j \neq i} w_{ij}\mathbf{p}_{ij} \otimes \mathbf{p}_{ij} \tag{3.72}$$

$$\mathbf{b}_i = \sum_{j \neq i} w_{ij}\mathbf{p}_{ij}\left(\phi_j - \phi_i\right) \tag{3.73}$$

If the moment matrix $\mathbf{M}_i$ is not singular, a unique solution exists, and the spatial derivatives are obtained as

$$\partial\phi_{\mathbf{r}=\mathbf{r}_i} = \mathbf{H}_{r_s}\mathbf{M}_i^{-1}\mathbf{b}_i \tag{3.74}$$

The above spatial derivatives can be rearranged into the following schemes:

$$\langle \nabla \phi \rangle_i = \sum_{j \neq i} \left\{ \frac{w_{ij}}{r_s} \left(\phi_j - \phi_i \right) \left(\begin{bmatrix} \mathbf{M}_1 \\ \mathbf{M}_2 \end{bmatrix} \mathbf{p}_{ij} \right) \right\} \qquad (3.75)$$

$$\langle \nabla \cdot \mathbf{u} \rangle_i = \sum_{j \neq i} \left\{ \frac{w_{ij}}{r_s} \left(\mathbf{u}_j - \mathbf{u}_i \right) \left(\begin{bmatrix} \mathbf{M}_1 \\ \mathbf{M}_2 \end{bmatrix} \mathbf{p}_{ij} \right) \right\} \qquad (3.76)$$

$$\langle \nabla^2 \phi \rangle_i = \sum_{j \neq i} \left\{ 2 \frac{w_{ij}}{r_s^2} \left(\phi_j - \phi_i \right) [\mathbf{M}_3 + \mathbf{M}_4] \cdot \mathbf{p}_{ij} \right\} \qquad (3.77)$$

where $\mathbf{M}_k$ represents the k-th row of matrix $\mathbf{M}_i^{-1}$.

In type B scheme, $\phi(\mathbf{r}_i)$ is treated as an unknown value. Thus a Taylor series expansion is given as

$$\phi_j = \sum_{m=0}^{p} \frac{1}{m!} (\mathbf{r}_{ij} \cdot \nabla)^m \phi|_{\mathbf{r}_i} + o\left(|\mathbf{r}_{ij}|^{p+1} \right) \qquad (3.78)$$

By ignoring higher-order terms the following equations can be obtained:

$$\phi_j = \sum_{m=0}^{p} \frac{r_s^m}{m!} \left(\frac{\mathbf{r}_{ij}}{r_s} \cdot \nabla \right)^m \phi|_{\mathbf{r}_i} = \hat{\mathbf{p}}_{ij} \cdot \left(\hat{\mathbf{H}}_{r_s}^{-1} \hat{\partial} \phi|_{\mathbf{r}_i} \right) \qquad (3.79)$$

where

$$\hat{\partial} = \begin{bmatrix} 1 \\ \partial \end{bmatrix} \qquad (3.80)$$

$$\hat{\mathbf{H}}_{r_s} = \begin{bmatrix} 1 & 0 \\ 0 & \mathbf{H}_{r_s} \end{bmatrix} \qquad (3.81)$$

$$\hat{\mathbf{p}}_{ij} = \hat{\mathbf{p}} \left(\frac{\mathbf{r}_{ij}}{r_s} \right) = \begin{bmatrix} 1 \\ \mathbf{p}_{ij} \end{bmatrix} \qquad (3.82)$$

For the second-order scheme ($p = 2$) in three dimensions, these functions are written as

$$\hat{\partial} = \left[1, \frac{\partial}{\partial x}, \frac{\partial}{\partial y}, \frac{\partial}{\partial z}, \frac{\partial^2}{\partial x^2}, \frac{\partial^2}{\partial y^2}, \frac{\partial^2}{\partial z^2}, \frac{\partial^2}{\partial x \partial y}, \frac{\partial^2}{\partial y \partial z}, \frac{\partial^2}{\partial x \partial z} \right]^T \qquad (3.83)$$

$$\hat{\mathbf{H}}_{r_s} = \operatorname{diag} \left(1, \frac{1}{r_s}, \frac{1}{r_s}, \frac{1}{r_s}, \frac{2}{r_s^2}, \frac{2}{r_s^2}, \frac{2}{r_s^2}, \frac{1}{r_s^2}, \frac{1}{r_s^2}, \frac{1}{r_s^2} \right) \qquad (3.84)$$

$$\hat{\mathbf{p}}_{ij} = \hat{\mathbf{p}}\left(\frac{\mathbf{r}_{ij}}{r_s}\right) = \left[1, \frac{x_{ij}}{r_s}, \frac{y_{ij}}{r_s}, \frac{z_{ij}}{r_s}, \frac{x_{ij}^2}{r_s^2}, \frac{y_{ij}^2}{r_s^2}, \frac{z_{ij}^2}{r_s^2}, \frac{x_{ij}y_{ij}}{r_s^2}, \frac{y_{ij}z_{ij}}{r_s^2}, \frac{x_{ij}z_{ij}}{r_s^2}\right]^T \quad (3.85)$$

Similarly the spatial derivatives in the type B scheme are obtained as

$$\hat{\partial}\phi_{\mathbf{r}=\mathbf{r}_i} = \hat{\mathbf{H}}_{r_s}\hat{\mathbf{M}}_i^{-1}\hat{\mathbf{b}}_i \quad (3.86)$$

with

$$\hat{\mathbf{M}}_i = \sum_{j\neq i} w_{ij}\hat{\mathbf{p}}_{ij} \otimes \hat{\mathbf{p}}_{ij} \quad (3.87)$$

$$\hat{\mathbf{b}}_i = \sum_{j\neq i} w_{ij}\hat{\mathbf{p}}_{ij}\phi_j \quad (3.88)$$

Compared with the classical discretization scheme, remarkable accuracy and high-order consistency can be achieved. As shown in Fig. 3.3 the convergence of the error concerning spatial resolution agrees with the scheme orders.

3.4 Conservative consistent MPS scheme

In previous sections, several improved schemes have been introduced. Although high-order accuracies are guaranteed in these improved schemes, they are not conservative. The original MPS is conservative due to the spatial symmetry of the kernel function. The spatial symmetry of the kernel function is broken after reconstruction in the corrective MPS [18] and LSMPS [17], and consequently the conservation property is violated. To ensure both high-order accuracy and conservativeness, Liu et al. [19] have developed a conservative consistent MPS scheme, called pairwise-relaxing method. In their schemes, pairwise-relaxing coefficients are introduced to the kernels to provide the degree of freedom to enforce the Taylor-series consistency condition while the antisymmetric structure of the scheme is maintained.

The original MPS scheme in conservative form is

$$\langle\nabla\phi\rangle_i = \frac{d}{n^0}\sum_{j\neq i}\frac{(\phi_j + \phi_i)}{r_{ij}}\mathbf{r}_{ij}w_{ij} \quad (3.89)$$

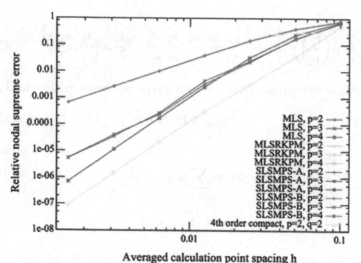

Averaged calculation point spacing h

Figure 3.3 Relative error by the LSMPS scheme [17].

To introduce an extra degree of freedom to enforce the consistency condition, pairwise relaxing coefficients c_{ij} are then introduced:

$$\langle \nabla \phi \rangle_i = \frac{d}{n^0} \sum_{j \neq i} c_{ij} \frac{(\phi_j + \phi_i)}{r_{ij}} \mathbf{r}_{ij} w_{ij} \qquad (3.90)$$

By considering Taylor-series expansion of ϕ_{ij} about ϕ_i, one obtains

$$\langle \nabla \phi \rangle_i = \frac{d}{n^0} \sum_{j \neq i} c_{ij} \frac{2\phi_i}{r_{ij}} \mathbf{r}_{ij} w_{ij} + \frac{d}{n^0} \sum_{j \neq i} c_{ij} \frac{(\frac{\partial \phi}{\partial x})_i x_{ij}}{r_{ij}} \mathbf{r}_{ij} w_{ij} + \frac{d}{n^0} \sum_{j \neq i} c_{ij} \frac{(\frac{\partial \phi}{\partial y})_i y_{ij}}{r_{ij}} \mathbf{r}_{ij} w_{ij} + O(\Delta l)$$

$$(3.91)$$

To ensure first-order accuracy the following equations should be satisfied:

$$\frac{d}{n_0} \sum_{j \neq i} c_{ij} w_{ij} \mathbf{X}_i / r_{ij}^2 = \mathbf{b}_i \qquad (3.92)$$

where $\mathbf{X}_i = (x_{ij}, y_{ij}, x_{ij}^2, y_{ij}^2, x_{ij} y_{ij})^T$ and $\mathbf{b}_i = (0, 0, 1, 1, 0)^T$.

Combining the consistency equations for all the particles and enforcing $c_{ij} = c_{ji}$ to keep the antisymmetric feature a set of equations in matrix form are obtained:

$$\mathbf{Ac} = \mathbf{b}$$
$$c_{ij} = c_{ji}, \qquad (3.93)$$

Here $\mathbf{A}$ is a $5N \times \sum_{i=1}^{N} p_i$ matrix, N is the number of particles and p_i is the interaction particle number for particle i, $\mathbf{c}$ is a $\sum_{i=1}^{N} p_i$-length vector that lists the relaxing coefficients for interaction particle pairs, and c_{ij} is chosen to minimize

$$J = \sum_{j \neq i} (c_{ij} - 1)^2, \qquad (3.94)$$

The total number of unknowns in Eq. (3.93) is $\sum_{i=1}^{N} p_i$, and the number of equations is $5N + \frac{\sum_{i=1}^{N} p_i}{2}$. To make the system underdetermined, it requires

$$\sum_{i=1}^{N} p_i > 10N. \qquad (3.95)$$

Eq. (3.95) can be satisfied easily by setting the interaction radius to make $p_i > 10$. The underdetermined system of equations consisting of Eqs. (3.93) and (3.94) can be solved by the LSMR algorithm proposed by Fong and Saunders [20].

3.5 Particle-mesh coupling method

Although MPS method has the advantage in automatic interface capture, it suffers from inaccuracy and high computation cost. On the other hand, traditional mesh method is accurate and efficient. To take advantage of both MPS method and mesh method the particle-mesh coupling method is introduced.

Liu et al. [21] proposed a hybrid method for multiphase flow. In their hybrid method, one phase is represented by moving particles and the other phase is defined on a stationary mesh (see Fig. 3.4). The flow field is discretized by a conservative finite volume approximation on the stationary mesh, and the interface is automatically captured by the distribution of particles moving through the stationary mesh. Governing equations are first solved on the background mesh using finite volume method. The governing equations solved by the finite volume method are

$$\oint_S \mathbf{v} \cdot \mathbf{n} dS = 0 \qquad (3.96)$$

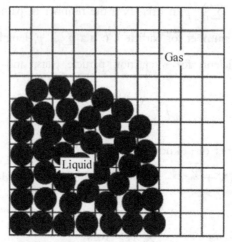

Figure 3.4 Setup of mesh and particle [21].

$$\frac{\partial}{\partial t}\int_V \rho \mathbf{v}dV + \oint_S \rho \mathbf{v}(\mathbf{v}\cdot\mathbf{n})dS = -\oint_S p\cdot\mathbf{I}\cdot\mathbf{n}dS - \oint_S \boldsymbol{\tau}\cdot\mathbf{n}dS + \int_V \mathbf{f}dV$$

(3.97)

Computational solutions are obtained on a staggered mesh shown in Fig. 3.4. In the staggered mesh the control volume used for conservation of mass is centered at the cell centroid, the one of the conservation of momentum in the x-direction is centered at the right face, and the one for the conservation of momentum in the y-direction is centered at the upper face, as shown in Fig. 3.5.

Since the fluid velocities are computed on the fixed mesh and particles move with the fluid velocities, the velocity of particles must be found by interpolating from the fixed mesh.

The transfer of variables between particles and meshes is the important operation of this method. Fig. 3.6 shows the extrapolation of surface force from particle to mesh. One should find the mesh where the center of the interface particle locates and then search the neighbor meshes that are occupied by the interface particle. The fractional areas that the interface particle occupy can be calculated by considering the particle as square shape in two dimensions and cubic shape in three dimensions. Finally the surface tension on the mesh can be calculated using area-weighted extrapolation, as follows:

$$(\vec{f_v})_k = \frac{1}{V}\vec{f_{sv}}A_k$$

(3.98)

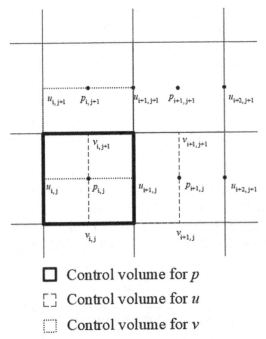

Figure 3.5 Staggered mesh in mesh calculation of hybrid method [21].

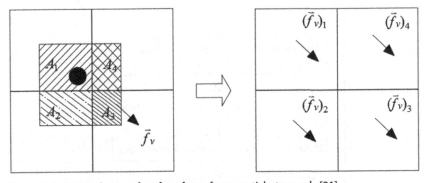

Figure 3.6 Extrapolation of surface force from particle to mesh [21].

where $(f_v^{\rightarrow})_k$ is the surface force at each mesh k, $f_{sv}^{\rightarrow}$ is the surface force at the surface particle, V is the volume of the mesh, and A_k is the fractional area associated with $\sum_{k=1}^{4} A_k = 1$. The area-weight is only one method of extrapolation. The other kind of weight function may be effective as well, such as the distance-weight.

Since the fluid velocities are calculated on the background mesh but the particles need to move with the fluid velocities, the velocities of particles must be calculated by interpolating from the meshes. The interpolation starts by identifying the meshes that are closest to the particle. The particle velocity is then interpolated by area-weighted interpolation which incorporates most of the features developed for the extrapolation of the surface force described above [21].

Once the velocity of each particle has been found the velocity and position are updated according to the scheme of original MPS method. The solution algorithm of the hybrid method is shown in Fig. 3.7. Simulation results of Rayleigh–Taylor instability by this hybrid method are shown in Fig. 3.8. The heavier fluid is represented by particles, and the heavier fluid falls downward when the calculation starts. The interface is clearly captured by particles during the falling process.

Ishii et al. [22] developed another hybrid method by coupling the cubic interpolated propagation (CIP) method with the MPS method. Compared

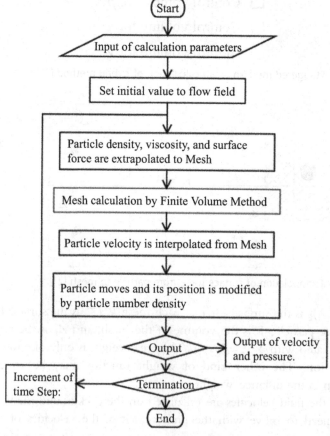

Figure 3.7 Solution algorithm of the hybrid method [21].

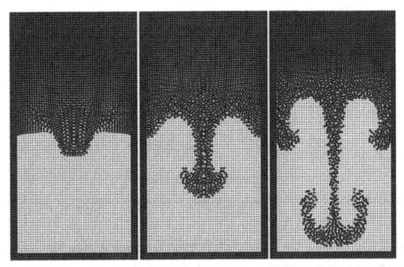

Figure 3.8 Simulation results of Rayleigh–Taylor instability by hybrid method ($t = 0.4$, 0.8, and 1.2) [21].

with the hybrid method of Liu et al. [21] where the whole liquid phase is represented by particles, in the hybrid method of Ishii et al. [22] the particles only distribute over the surface of the liquid phase. The bulk flow is solved using CIP FVM method, and MPS particles are used to reconstruct the interface by using the volume fraction of liquid that is calculated with CIP, as shown in Fig. 3.9. In each time step, particle insertion and deletion are permissible in accordance with the interface conditions of gas–liquid flow. In the hybrid method the velocities are interpolated and extrapolated between the particle coordinates and grid coordinates with weight functions. Velocities given by the advection part of MPS are combined with that given by the advection part of CIP within the hybrid method.

The velocity $\mathbf{u}_{MPS}$ given by the advection part of the MPS method is combined with the velocity $\mathbf{u}_{(CIP)}$ that is calculated by CIP but transferred to the particle coordinates:

$$\mathbf{u}^*_{MPS} = (1 - \lambda_{MPS})\mathbf{u}_{MPS} + \lambda_{MPS}\mathbf{u}_{(CIP)} \qquad (3.99)$$

where λ_{MPS} is written as

$$\lambda_{MPS} = \begin{cases} 0 \leftarrow \text{gas} & (\theta_{(CIP)} < \theta_{min}) \\ (\theta_{(CIP)} - \theta_{min})/(\theta_{max} - \theta_{min}) & (\theta_{min} \leq \theta_{(CIP)} \leq \theta_{max}) \\ 1 \leftarrow \text{liquid} & (\theta_{(CIP)} > \theta_{max}) \end{cases} \qquad (3.100)$$

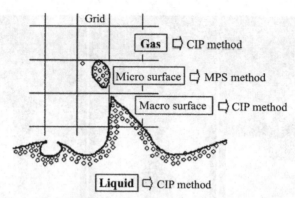

Figure 3.9 Assignment of CIP and MPS method to predict interfaces in gas–liquid flows [22].

where λ_{MPS} changes from 0 to 1 between gas and liquid regions. The variable $\theta_{(CIP)}$ is the volume fraction of the liquid phase calculated in the particle coordinates by

$$\theta_{(CIP)} = \frac{\sum_{i \neq j} \theta_{CIP} w(R_{ij})}{\sum_{i \neq j} w(R_{ij})} \qquad (3.101)$$

where θ_{CIP} is the volume fraction of the liquid phase calculated using CIP in terms of the grid coordinates, and θ_{min} and θ_{max} are usually set as 0.2 and 0.8, respectively.

Within CIP the velocity $\mathbf{u}_{CIP}$ given by the advection part is combined with velocity $\mathbf{u}_{(MPS)}$, which is calculated by MPS but transferred to the grid coordinates:

$$\mathbf{u}^*_{CIP} = (1 - \lambda_{CIP})\mathbf{u}_{CIP} + \lambda_{CIP}\mathbf{u}_{(MPS)} \qquad (3.102)$$

where λ_{CIP} is written as

$$\lambda_{CIP} = (1 - \alpha)\beta \qquad (3.103)$$

where

$$\alpha = \begin{cases} 0 \leftarrow \text{gas} & (\theta_{CIP} < \theta_{min}) \\ (\theta_{CIP} - \theta_{min})/(\theta_{max} - \theta_{min}) & (\theta_{min} \leq \theta_{CIP} \leq \theta_{max}) \\ 1 \leftarrow \text{liquid} & (\theta_{CIP} > \theta_{max}) \end{cases} \qquad (3.104)$$

and

$$\beta = \frac{\sum_{i \neq j} w(R_{ij})}{N_0} \qquad (3.105)$$

where N_0 is the PND in liquid, and as a result, β changes from 0 to 1.

Fig. 3.10 shows the calculation algorithm of CIP-MPS hybrid method. The advection part of CIP is solved by interpolating the spatial distributions of variables within a cell with a cubic polynomial function. Velocities in the advection part of the MPS method and CIP are

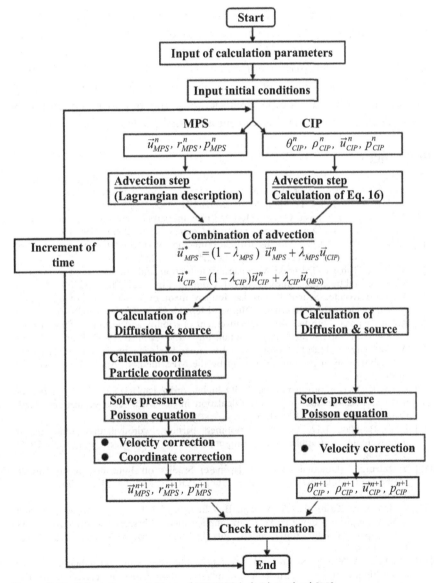

Figure 3.10 Calculation algorithm of CIP-MPS hybrid method [22].

combined in each time step; that is, the velocities in terms of both particle
and grid coordinates are modified by Eqs. (3.99) and (3.102). Thereafter
the diffusion term, external force term, and PPE are calculated with CIP
and MPS methods, respectively, according to their own algorithms. The
velocities, particle positions, and liquid volume fractions are then updated
at a new time step [22].

Although Ishii et al. [22] realized CIP-MPS hybrid calculation, the
algorithm shown by Fig. 3.10 is not coupled tightly between mesh and
particle methods. It will lead to numerical inconsistency in fluid properties. For instance the mass is not conserved [23]. Liu et al. [23] addressed
the problem of mass nonconservation by a volume-fraction remedy procedure and enforced a proper treatment for the boundary condition in
MPS calculation by using the concept of virtual particles.

References

[1] Hirokazu Ikeda, Seiichi Koshizuka, Yoshiaki Oka, Hyun Sun Park, Jun Sugimoto, Numerical analysis of jet injection behavior for fuel-coolant interaction using particle method, J. Nucl. Sci. Technol. 38 (3) (2001) 174−182.
[2] Abbas Khayyer, Hitoshi Gotoh, Modified moving particle semi-implicit methods for the prediction of 2D wave impact pressure, Coast. Eng. 56 (2009) 419−440.
[3] S. Koshizuka, Y. Oka, Moving-particle semi-implicit method for fragmentation of incompressible fluid, Nucl. Sci. Eng. 123 (1996) 421−434.
[4] Masayuki Tanaka, Takayuki Masunaga, Stabilization and smoothing of pressure in MPS method by quasi-compressibility, J. Comput. Phys. 229 (2010) 4279−4290.
[5] Masahiro Kondo, Seiichi Koshizuka, Improvement of stability in moving particle semi-implicit method, Internat, J. Numer. Methods Fluids 65 (2011) 638−654.
[6] Abbas Khayyer, Hitoshi Gotoh, Enhancement of stability and accuracy of the moving particle semi-implicit method, J. Comput. Phys. 230 (2011) 3093−3118.
[7] Abbas Khayyer, Hitoshi Gotoh, A higher order Laplacian model for enhancement and stabilization of pressure calculation by the MPS method, Appl. Ocean. Res. 32 (2010) 124−131.
[8] Abbas Khayyer, Hitoshi Gotoh, A 3D higher order Laplacian model for enhance-ment and stabilization of pressure calculation in 3D MPS-based simulations, Appl. Ocean. Res. 37 (2012) 120−126.
[9] P.W. Randles, L.D. Libersky, Smoothed particle hydrodynamics: some recent improvements and applications, Comput. Methods Appl. Mech. Eng. 139 (1996) 375−408.
[10] Y. Suzuki, (Doctoral Thesis in Japanese) Studies on Improvement in Particle Methodand Multiphysics Simulator Using Particle Method, The University of Tokyo, 2008.
[11] G. Duan, S. Koshizuka, A. Yamaji, B. Chen, X. Li, T. Tamai, An accurate and stable multiphase moving particle semi-implicit method based on a corrective matrix for all particle interaction models, Int. J. Numer. Methods Eng. 115 (2018) 1287−1314.
[12] G. Duan, A. Yamaji, S. Koshizuka, B. Chen, The truncation and stabilization error in multiphase moving particle semi-implicit method based on corrective matrix: which is dominant, Comput. Fluids. 190 (2019) 254−273.

[13] L. Brookshaw, A method of calculating radiative heat diffusion in particle simulations, Publ. Astron. Soc. Aust. 6 (1985) 207–210.

[14] G. Duan, T. Matsunaga, A. Yamaji, S. Koshizuka, M. Sakai, Imposing accurate wall boundary conditions in corrective-matrix-based moving particle semi-implicit method for free surface flow, Int. J. Numer. Methods Fluids. 93 (2021) 148–175.

[15] T. Tamai, S. Koshizuka, Least squares moving particle semi-implicit method: an arbitrary high order accurate meshfree Lagrangian approach for incompressible flow with free surfaces, Comput. Part. Mech. 1 (2014) 441.

[16] G. Duan, T. Matsunaga, S. Koshizuka, A. Yamaguchi, M. Sakai, New insights into error accumulation due to biased particle distribution in semi-implicit particle methods, Comput. Methods Appl. Mech. Eng. (2021). Under Review.

[17] T. Tamai, S. Koshizuka, Least squares moving particle semi-implicit method, Comput. Part. Mech. 1 (3) (2014) 277–305.

[18] A. Khayyer, H. Gotoh, Enhancement of stability and accuracy of the moving particle semi-implicit method, J. Comput. Phys. 230 (8) (2011) 3093–3118.

[19] X. Liu, K. Morita, S. Zhang, An ALE pairwise-relaxing meshless method for compressible flows, J. Comput. Phys. 387 (15) (2019) 1–13.

[20] D.C.L. Fong, M. Saunders, LSMR: an iterative algorithm for sparse least-squares problems, SIAM J. Sci. Comput. 33 (5) (2011) 2950–2971.

[21] J. Liu, S. Koshizuka, Y. Oka, A hybrid particle-mesh method for viscous, incompressible, multiphase flows, J. Comput. Phys. 202 (1) (2005) 65–93.

[22] E. Ishii, T. Ishikawa, Y. Tanabe, Hybrid particle/grid method for predicting motion of micro-and macrofree surfaces, J. Fluids Eng. 128 (2006) 921–930.

[23] X. Liu, K. Morita, S. Zhang, A conservative finite volume-particle hybrid method for simulation of incompressible interfacial flow, Comput. Method. Appl. M. 355 (2019) 840–859.

CHAPTER FOUR

Stabilization methods

A numerical method is said to be stable if the errors that appear in the solution process are not magnified continuously or rapidly. Otherwise the numerical method is said to suffer from an instability. In this perspective, stability analysis is difficult in Lagrangian particle methods. Particularly the stability analysis is rather challenging for the moving particle semi-implicit (MPS) method because the pressure field is solved implicitly and globally. Several kinds of instabilities in the explicit smooth particle hydrodynamics (SPH), like tensile instability [1] and pairing instability [2], have been investigated thoroughly. However, these studies are based on the conservative models and the explicit algorithm.

In MPS, Koshizuka and Oka [3] first pointed out that maintaining the repulsive forces in the pressure gradient model could effectively produce stable simulations. Khayyer and Gotoh [4] analyzed the stability of the original MPS models based on the pairwise repulsive and attractive forces. They argued that the MPS method seemed more prone to (tensile) instability than SPH [4]. Recently, Duan et al. [5] presented an in-depth stability analysis for the discretization models based on the variable differences (like the MPS models). They argued that biased neighbor support is the main reason for the fast error growth that can easily trigger instability [5]. Biased neighbor support means that the contribution of neighboring particles in a specific direction is significantly larger than that in the other directions. Therefore maintaining the unbiased/homogeneous neighbor supports for the internal particles plays an essential role in keeping the simulations stable.

Based on the above stability analysis the key points to maintain the stability of particle methods are as follows: (1) the velocity/pressure fluctuations must be effectively suppressed and (2) the isotropic or homogeneous particle distributions must be maintained. In this chapter, various adjustment techniques are presented to achieve the above two points. They can be roughly clarified into the velocity-adjusting techniques (see Sections 4.1, 4.3–4.5) and position-adjusting techniques (see Section 4.2 and 4.3).

Moving Particle Semi-implicit Method.
DOI: https://doi.org/10.1016/B978-0-443-13508-8.00004-4

4.1 Gradient model with stabilizing force

In the original MPS method the stability is maintained by modifying the pressure gradient models. The original zero–order gradient model is as follows:

$$\langle \nabla p \rangle_i = \frac{d}{n^0} \sum_{j \neq i} \frac{p_j - p_i}{r_{ij}} \frac{(\mathbf{r}_j - \mathbf{r}_i)}{r_{ij}} w_{ij} \tag{4.1}$$

where d is the dimensional number, n^0 is the constant particle number density (PND), r_{ij} is the particle distance, and w_{ij} is the weight function. It must be noted that the instability happens easily when the above model is directly employed. In the standard MPS method, Koshizuka and Oka [3,6] proposed the following pressure gradient model to keep the simulations stable:

$$\langle \nabla p \rangle_i = \frac{d}{n^0} \sum_{j \neq i} \frac{p_j - p_{i,min}}{r_{ij}} \frac{(\mathbf{r}_j - \mathbf{r}_i)}{r_{ij}} w_{ij} \tag{4.2}$$

where $p_{i,min}$ is the minimal pressure among the neighboring particle j and the center particle i. Because $p_j - p_{i,min}$ always holds, the physical repulsive pressure-gradient forces among two particles are guaranteed. Therefore the simulations become stable. The stability mechanism of Eq. (4.2) can also be interpreted in the following manner. Specifically, the above model can also be written as

$$\langle \nabla p \rangle_i = \frac{d}{n^0} \sum_{j \neq i} \frac{p_j - p_i}{r_{ij}} \frac{(\mathbf{r}_j - \mathbf{r}_i)}{r_{ij}} w_{ij} + \frac{p_i - p_{i,min}}{n^0} \sum_{j \neq i} \frac{d}{r_{ij}} \frac{(\mathbf{r}_j - \mathbf{r}_i)}{r_{ij}} w_{ij} \tag{4.3}$$

The last term of Eq. (4.3) plays a key role in keeping stability, and thus, it can be called a particle stabilizing term [7,8]. Because $p_i - p_{i,min}$ is always positive, this term can produce an adjusting force that always drives particles to move from the particle dense area to the particle diluted area. This can help to make the particle distributions homogeneous. As mentioned at the beginning of this chapter the homogeneous particles can effectively enhance the simulation stability.

It must be noted that the performance of stability techniques seriously depends on whether a particle is near a free surface or not. In other words, some stabilization techniques may fail near free surfaces. When the stabilization techniques are discussed the fluid particles can be classified into three groups: free-surface, (free-surface) vicinity, and internal particles, as shown in Fig. 4.1. Vicinity particles are defined as the particles of which the closest distance to the free particles is less than the interaction radius, r_e.

Based on the particle clarification in Fig. 4.1, it is stressed that the modified model, Eq. (4.2), can only be applied to internal particles. As shown in Fig. 4.2A the particle stabilizing term in Eq. (4.3) is usually minor for internal particles, and it can help to make the particle distributions homogeneous in this situation. On the other hand, as shown in Fig. 4.2B, this term is large near a free surface, and it always points to the outside of the free surface. In this situation, when $p_i - p_{i,min}$ is large (e.g., $p_{i,min}$ is a negative value) the particle stabilizing term can easily overadjust the particle velocity, causing the particles to fly away from the free surfaces (i.e., numerical splash).

Nevertheless, a simple modification can make the model in Eq. (4.2) applicable to the vicinity and free-surface particles. Specifically the negative pressure under free surfaces must be set as zero to avoid the possible over adjustments due to $p_{i,min}$. When this special modification is employed, the gradient model in Eq. (4.2) can be applied to all the free-surface, vicinity, and internal particles. The reason is discussed as follows: For free-surface particles, both p_i and $p_{i,min}$ are zeros. The particle stabilizing term in Eq. (4.3) automatically disappears for the free-surface particles. For the vicinity particles, p_i is not zero, but $p_{i,min}$ is always zero. In this situation the particle stabilizing term is proportional to p_i. For normal free-surface flows the pressure p_i decreases to zero from the inside to free

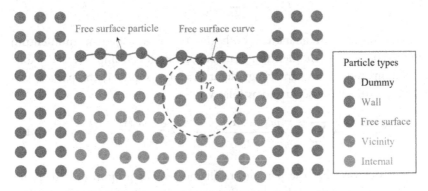

Figure 4.1 Particle clarification for the stabilization techniques [5].

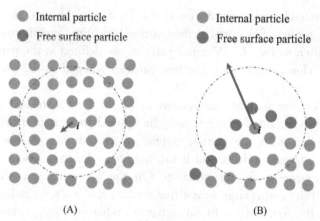

Figure 4.2 Sketch showing the particle stabilizing term supplemented into the pressure gradient model. (A) For internal particles [9] and (B) near a free surface.

surfaces, indicating an automatically decreasing particle stabilizing term [10]. Thus the particle stabilizing term can provide reasonable adjustment in the surface normal direction, without causing severe numerical splash [10].

In brief, to make the pressure gradient model in Eq. (4.2) work well, one must set the negative pressure under free surfaces to zero. In this situation, there is no adjustment for the free-surface particles. The collision model in Section 4.3 is recommended to further enhance the stability at the free surface.

The second stable pressure gradient model is the conservative model [11] as follows:

$$\langle \nabla p \rangle_i = \frac{d}{n^0} \sum_{j \neq i} \frac{p_j + p_i}{r_{ij}} \frac{(\mathbf{r}_j - \mathbf{r}_i)}{r_{ij}} w_{ij} \tag{4.4}$$

This model can guarantee both the linear and angular momentum conservations. The momentum conservation itself can improve the stability. Compared to the standard gradient model in Eq. (4.1) the particle stabilizing term can be separated as follows:

$$\langle \nabla p \rangle_i = \frac{d}{n^0} \sum_{j \neq i} \frac{p_j - p_i}{r_{ij}} \frac{(\mathbf{r}_j - \mathbf{r}_i)}{r_{ij}} w_{ij} + \frac{2p_i}{n^0} \sum_{j \neq i} \frac{d}{r_{ij}} \frac{(\mathbf{r}_j - \mathbf{r}_i)}{r_{ij}} w_{ij} \tag{4.5}$$

It is obvious that the particle stabilizing term is proportional to p_i for all the fluid particles. Therefore this model is applicable to all the fluid particles including the free-surface and vicinity particles. However, it must be noted that this model is not suitable for the negative pressure field. For the internal particles with positive pressure the magnitude of the particle stabilizing term in Eq. (4.5) is usually larger than that in Eq. (4.3). Due to this reason, the numerical diffusion of the model in Eq. (4.4) is much stronger than that in Eq. (4.2) [8,10,12], even though the stability is improved further.

Another stable pressure gradient model is proposed by Shibata et al. [13] as follows:

$$\langle \nabla p \rangle_i = \frac{d}{n^0} \sum_{j \neq i} \frac{p_j}{r_{ij}} \frac{(\mathbf{r}_j - \mathbf{r}_i)}{r_{ij}} w_{ij} \tag{4.6}$$

Similarly the particle stabilizing term in this model can be decomposed as follows:

$$\langle \nabla p \rangle_i = \frac{d}{n^0} \sum_{j \neq i} \frac{p_j - p_i}{r_{ij}} \frac{(\mathbf{r}_j - \mathbf{r}_i)}{r_{ij}} w_{ij} + \frac{p_i}{n^0} \sum_{j \neq i} \frac{d}{r_{ij}} \frac{(\mathbf{r}_j - \mathbf{r}_i)}{r_{ij}} w_{ij} \tag{4.7}$$

For example the magnitude of the particle stabilizing term in Eq. (4.7) is just a half of that in Eq. (4.5). This model is also suitable for all the internal, vicinity, and free-surface particles. It was reported that this model could even provide stable results for the problem with negative pressure by using the virtual particle method [13]. For the internal particles with normal positive pressure the adjustment magnitude by Eq. (4.7) is still probably larger than that in the original stable pressure gradient model in Eq. (4.3). It is noted that the particle stabilizing term in Eq. (4.3) near a free surface is actually simplified to that in Eq. (4.7).

Up to now, we have two main principles for the design of the particle stabilizing terms in MPS as follows:

1. *The coefficient pressure for the internal particles must be positive. In this manner the particle can always move from the particle dense region to the particle diluted region, helping to keep the particle distribution homogeneous.*
2. *The coefficient pressure for the free-surface and vicinity particles must be proportional to the center pressure p_i. In this manner the adjustment magnitude can automatically decrease when a particle is approaching the free surface.*

Based on the above principles, it is possible to design more advanced stable and accurate pressure gradient models for particles. For example,

Duan et al. [8] proposed the following pressure gradient model for the MPS method:

$$\langle \nabla p \rangle_i = \begin{cases} \dfrac{d}{n^0} \displaystyle\sum_{j \neq i} \dfrac{p_j + p_i}{r_{ij}} \dfrac{(\mathbf{r}_j - \mathbf{r}_i)}{r_{ij}} w_{ij} & i \in \mathbb{F} \text{ or } \mathbb{V} \\[3ex] \dfrac{d}{n^0} \displaystyle\sum_{j \neq i} \dfrac{\left(p_j - p_i + \xi\left(p_{i,max} - p_{i,min}\right)\right)}{r_{ij}} \dfrac{(\mathbf{r}_j - \mathbf{r}_i)}{r_{ij}} w_{ij} & i \in \mathbb{I} \end{cases}$$

(4.8)

where $\mathbb{F}$, $\mathbb{V}$, and $\mathbb{I}$ denote the free-surface, vicinity, and internal particles, respectively, as shown in Fig. 4.1; ξ is a tuning parameter ($\xi = 0.2 \text{ to } 0.5$ is found suitable [8]). Specifically, the conservative pressure gradient model is used to guarantee the stability for the free-surface and vicinity particles, which take the incomplete neighbor support. On the other hand, the particle stabilizing term for the internal particles is proportional to the difference between the maximum and minimum pressures in the neighborhood. As shown in Fig. 4.3, when an unexpected low or high pressure happens the magnitude of the particle stabilizing term can be automatically increased. The produced larger adjustment can help to reduce the fluctuation and enhance stability. It is noted that only the low pressure can enhance the particle stabilizing term in the original model in Eq. (4.3). For example the high pressure was not considered in the particle stabilizing term. This drawback is overcome by the improved model in Eq. (4.8). Briefly the advanced model in Eq. (4.8) can deal with the

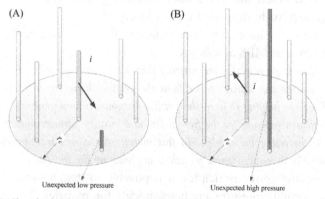

Figure 4.3 Sketch showing the unexpected (A) low pressure and (B) high pressure in the neighbor area due to pressure oscillation [8].

negative pressure of internal particles but might suffer instability when the negative pressure takes place at a free surface. The above idea to specify different gradient models based on whether the particle is near a free surface or not is quite important, especially when one would like to improve the simulation stability.

4.2 Particle shifting scheme

One main reason why the particle stabilizing term can produce stable simulation is that the adjusting force can drive particles to move from the particle dense region to the particle diluted region. Therefore the particle distribution becomes homogeneous. Recently, Duan et al. [5] demonstrated that the biased particle distribution is the main reason for the fast error accumulation and instability. In other words, maintaining the particle distribution homogeneous and unbiased can effectively keep the simulation stable. Thus one main purpose of the adjustment term is to maintain the particle distribution always homogeneous. In this regard, it is possible to directly shift particles to maintain the distribution homogeneous. This is the basic idea of the particle shifting (PS) technique [14]. Because the PS technique does not significantly affect the velocity field, it can usually produce more accurate results. Furthermore the influence of PS can be considered by interpolating the velocity at the new positions based on the Taylor series expansion [14].

4.2.1 Original PS technique for internal particles

The PS technique is first proposed by Xu et al. [14] for the SPH method. Like the particle stabilizing term in particle methods, PS directly shifts the particles from the particle dense region to the particle diluted region.

For the MPS method the following shifting method from Ref. [10] can be adopted:

$$\delta \mathbf{r}_i = \frac{a l_0}{n^0} \sum_{j \neq i} \frac{l_0}{r_{ij}} \frac{\left(\mathbf{r}_i - \mathbf{r}_j\right)}{r_{ij}} w_{ij} \tag{4.9}$$

where a is a coefficient ($a = 0.15$ is found suitable for most simulations). The dynamic way to specify the coefficient a can be found in Ref. [10]. It is noted that the shifting vector in Eq. (4.9) is rather similar to the particle stabilizing term in Eq. (4.3). This shifting vector always points to the particle diluted direction, as shown in Fig. 4.4. Therefore the PS technique can effectively maintain the particle distribution basically isotropic for the internal particles. Based on the isotropic particle distributions the accuracy of the discretization models can be significantly enhanced. Particularly, after PS is employed the consistent high-order schemes can produce the stable results. In this situation the accuracy can be effectively enhanced, as demonstrated in Refs. [10,15−17].

4.2.2 Surface tangential shifting for free-surface particles

Like the particle stabilizing term the PS technique also faces difficulty near the free surface. For example the shifting vector is too large near the free surface, easily causing numerical dilution. In this situation, Lind et al. [18] proposed to effectively reduce the shifting vector near free surface. A tuning parameter must be specified near free surfaces for the success of particle shifting. Then, Khayyer et al. [19] proposed an optimized particle shifting (OPS) technique to perform the particle shifting at free surfaces. The basic idea is as follows: As shown in Fig. 4.5, $\delta \mathbf{r}$ denotes the original PS vector calculated from Eq. (4.9), $\mathbf{n}$ denotes a surface normal, and $\delta \mathbf{r}'$ is the OPS shifting vector. For example the shifting component in the surface tangential direction can be kept, but the shifting component in the surface normal direction must be deleted.

Specifically the OPS vector can be calculated from

$$\delta \mathbf{r}'_i = \delta \mathbf{r}_i - (\delta \mathbf{r}_i \cdot \mathbf{n})\mathbf{n} \qquad (4.10)$$

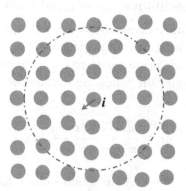

Figure 4.4 Sketch showing the particle shifting vector which always points to the particle diluted direction [9].

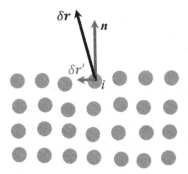

Figure 4.5 Sketch showing the optimized particle shifting technique for the surface tangential shifting at a free surface.

where **n** is another surface normal at the free surface. As shown in Fig. 4.5, δr is in a surface normal direction. To make the OPS work the surface normal **n** must be in a different direction from δr. There are two main methods to calculate the surface normal **n**.

First a more accurate surface normal can be calculated based on a corrective matrix, as proposed by Khayyer et al. [19]. For example the surface normal is calculated as follows:

$$\mathbf{n} = \frac{\mathbf{N}}{\|\mathbf{N}\|}, \mathbf{N} = \frac{d}{n^0} \sum_{j \neq i} \frac{1}{r_{ij}} \mathbf{C}_i \frac{(\mathbf{r}_i - \mathbf{r}_j)}{r_{ij}} w_{ij} \qquad (4.11)$$

where the corrective matrix $\mathbf{C}_i$ is calculated in the following manner:

$$\mathbf{C}_i^{-1} = \frac{d}{n^0} \begin{bmatrix} \sum_{j \neq i} \frac{x_{ij}^2}{r_{ij}^2} w_{ij} & \sum_{j \neq i} \frac{x_{ij} y_{ij}}{r_{ij}^2} w_{ij} \\ \sum_{j \neq i} \frac{y_{ij} x_{ij}}{r_{ij}^2} w_{ij} & \sum_{j \neq i} \frac{y_{ij}^2}{r_{ij}^2} w_{ij} \end{bmatrix} \qquad (4.12)$$

It is noted that the original OPS is developed in the SPH framework. Some adjustments, especially the calculation of **n**, are applied above in the MPS framework. Khayyer et al. [19] argued that the corrective matrix can improve the accuracy of surface normal and thus that the remaining tangential shifting vector can always shift the particle in the correct direction.

Second, there is a more simple yet effective method to calculate the surface normal **n**, as proposed by Duan et al. [10]. Specifically the surface normal can be simply calculated as follows:

$$\mathbf{n} = \frac{\mathbf{N}}{\|\mathbf{N}\|}, \mathbf{N} = \frac{d}{n^0} \sum_{j \neq i} \frac{(\mathbf{r}_i - \mathbf{r}_j)}{r_{ij}} w_{ij} \qquad (4.13)$$

Duan et al. [10] argued that the weight function to calculate the surface normal plays a key role to make the OPS work well. When the weight function is less sensitive to the particle approaching than the intrinsic weight function in calculating the original $\delta\mathbf{r}'$, the resultant tangential component $\delta\mathbf{r}'$ always points to the particle diluted direction in the surface tangential direction, as shown in Fig. 4.5. Therefore $\delta\mathbf{r}'$ always tries to make the free-surface particles evenly spaced at a free surface.

Finally, it is noted that the OPS techniques were applied to both free-surface and vicinity particles in the original studies [10,19]. However, the thickness of the vicinity-particle layers is another tuning parameter. Because (1) OPS works better for the free-surface particles than the vicinity particles and (2) there is a better shifting technique for the vicinity particles, it is recommended here that OPS is only applied to the free-surface particles. The corresponding strategy is elaborated below.

4.2.3 Improved particle shifting for vicinity particles

Wang et al. [20] proposed an improved particle shifting technology (IPST) for the particles near a free surface. This technique is particularly suitable for the vicinity particles. The idea is relatively simple, and the technique is effective. As shown in Fig. 4.6, to avoid the overshifting for the vicinity particles, it is necessary to reduce the interaction radius for PS. For a vicinity particle i, the particle distance to the closest free-surface particle, r_{fi} is computed first. Subsequently, the shifting vector can be calculated as follows [5,9]:

$$\delta\mathbf{r}_i'' = \frac{a l_0}{n^0} \sum_{(j \neq i \text{ and } r_{ij} < (r_f + 0.1 l_0))} \left\{ \frac{l_0}{r_{ij}} \frac{\mathbf{r}_{ij}}{r_{ij}} w_{ij} \right\} \qquad (4.14)$$

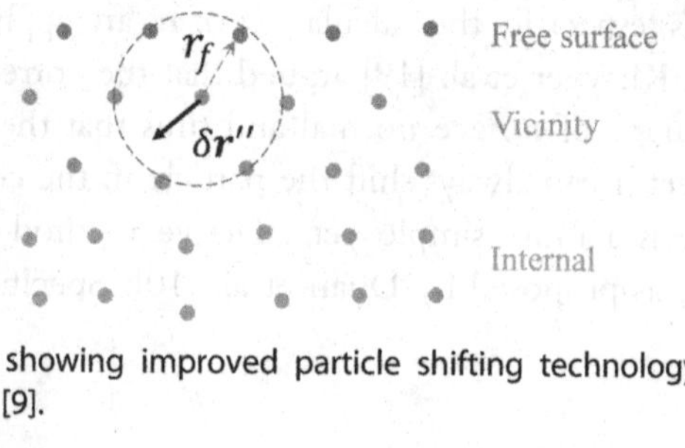

Figure 4.6 Sketch showing improved particle shifting technology for the particles near a free surface [9].

where a is the same coefficient for the original PS in Eq. (4.9). It is noted that this shifting vector is quite like the original one in Eq. (4.9). The only difference is that the interaction radius is reduced from r_e to r_f. It is also noted that based on this shifting vector a vicinity particle can always be moved to the diluted direction in the local area of r_f. Meanwhile, $\delta\mathbf{r}_i''$ can effectively prevent the particle clumping near a free surface.

4.2.4 Surface normal shifting for free-surface particles

One shortcoming of the above PS models is that the shifting vector in the surface normal direction is never considered for free-surface particles. In numerical simulations, the position fluctuations of the free-surface particles can happen. This kind of fluctuation can further trigger velocity and pressure fluctuations near free surfaces [5,17]. On the other hand, the shifting in the surface normal direction is challenging because such shifting can easily change the physical deformation of free surfaces.

Recently, Duan et al. [9] proposed a surface normal particle shifting (SNPS) technique to bridge this gap. As shown in Fig. 4.7 the basic idea of SNPS is to shift the free-surface particles to a smooth reconstructed free surface. Specifically the procedures are as follows: First the free-surface particles must be precisely detected. Second the surface normal vector is accurately calculated. Third the coordinate transformation is performed for a reference free-surface particle and its neighboring

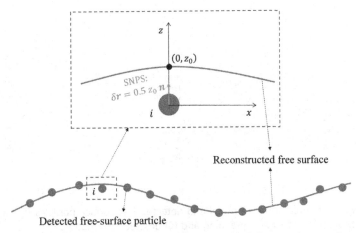

Figure 4.7 Sketch showing the surface normal particle shifting technique for the surface normal shifting at a free surface [9].

free-surface particles. In the transformed coordinate system the target z-axis is always in the surface normal direction, as shown in Fig. 4.7. The coordinate transformation can greatly reduce the difficulty of curve fitting in the next step. Fourth the curve fitting based on the LS method is employed to reconstruct the real free-surface boundary, as shown by the solid line in Fig. 4.7. Finally the distance from the reference free-surface particle to the reconstructed free-surface boundary is calculated, and the reference particle is gradually shifted exactly to this reconstructed boundary. The implementation details can be found in Ref. [9]. Even though the implementation procedures are complex, SNPS can produce rather smooth and high-quality free-surface boundaries for the droplet dynamics, as presented in Fig. 4.8. Last, it is noted that the SNPS technique can work well for the droplet flows, while it may smear the sharp angles at free surfaces for the large-scale free-surface flows.

Finally the applicability of different PS techniques is summarized below. As shown in Fig. 4.9 the original PS technique can only be applied to the internal particles. IPST can be applied to the vicinity particles. For the free-surface particles the OPS technique can provide the surface-tangential shifting vector, and the SNPS technique can provide the surface-normal shifting vector. It is noted that the SNPS technique is specially designed for the free-surface flow with surface tension. All the above techniques can be applied to both 2D and 3D. The combination of these PS techniques in Fig. 4.9 are employed in Refs. [5,9].

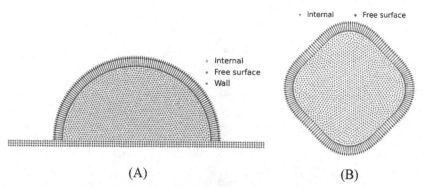

(A)   (B)

Figure 4.8 Calculated surface normal and detected free-surface particles in different droplet flows [9]. (A) Droplet spreading and (B) droplet deformation.

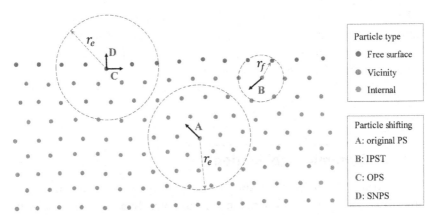

Figure 4.9 Summary of the particle shifting techniques for different particles [9].

4.3 Collision model

In the MPS method, when a pair of particles become quite close to each other, the weight function will increase to a rather large value. Then the interaction force between the two particles can be much larger than other particle pairs. For example the neighbor support will be severely biased, and fast error accumulation (i.e., instability) can happen easily [5]. Therefore the particle clumping must be avoided. The collision model is designed to separate two close particles.

The collision model was presented in Ref. [21]. The basic idea is to add additional repulsive force when two particles get close to each other. For a reference particle i and a neighboring particle j the collision model will be used (1) when the particle distance r_{ij} is smaller than al_0 and (2) the approaching velocity $(\mathbf{u}_j - \mathbf{u}_i)(\mathbf{r}_j - \mathbf{r}_i)/r_{ij}$ is less than zero. The typical value of a is 0.5 to 0.9. Then the new velocity of particle i after the collision model can be calculated from

$$\mathbf{u}'_i = \mathbf{u}_i + \frac{b(\mathbf{u}_j - \mathbf{u}_i)(\mathbf{r}_j - \mathbf{r}_i)}{r_{ij}} \frac{\mathbf{r}_j - \mathbf{r}_i}{r_{ij}} \quad (4.15)$$

where b is a scaling parameter. The typical value of b is 0.1 to 0.25. Meanwhile the new position of particle i after the collision can be calculated from

$$\mathbf{r}'_i = \mathbf{r}_i - cl_0 \frac{\mathbf{r}_j - \mathbf{r}_i}{r_{ij}} \quad (4.16)$$

where c is an adjusting parameter. The typical value of c is 0.2.

Last, it must be noted that even though the collision model can improve the stability, it will cause additional numerical diffusion. Besides, it cannot prevent the large void gaps among the internal particles. These are the main drawbacks of the collision model.

4.4 Dynamic stabilization

Dynamic stabilization was proposed by Tsuruta et al. [22] in the MPS framework. The dynamic stabilization aims to precisely provide the adequate repulsive interparticle forces based on the instantaneous distribution of particles to eliminate the possible particle penetration [22]. Specifically the pressure gradient model containing the dynamic stabilization term is as follows:

$$\langle \nabla p \rangle_i = \frac{d}{n^0} \sum_{j \neq i} \frac{p_j - p_i}{r_{ij}} \frac{(\mathbf{r}_j - \mathbf{r}_i)}{r_{ij}} w_{ij} + \frac{1}{n^0} \sum_{j \neq i} \mathbf{F}_{ij}^{DS} w_{ij} \qquad (4.17)$$

where the first term of the right hand side (RHS) is the standard pressure gradient model and the second term of RHS is the dynamic stabilization term. The repulsive force from a neighboring particle j to the target particle i, $\mathbf{F}_{ij}^{DS}$, is calculated from

$$\mathbf{F}_{ij}^{DS} = \begin{cases} 0 & r_{ij}^* \geq d_{ij} \\ -\rho_i \prod_{ij} \frac{\mathbf{r}_{ij}}{r_{ij}} r_{ij}^* & r_{ij}^* < d_{ij} \end{cases} \qquad (4.18)$$

where d is the particle diameter, the superscript $*$ indicates a state after the advection by the standard pressure gradient model, and $\prod_{ij}$ is a parameter to adjust magnitude of $\mathbf{F}_{ij}^{DS}$. The parameter d_{ij} is calculated as follows:

$$d_{ij} = (1 - \alpha_{\Delta t}) \frac{d_i + d_j}{2} \qquad (4.19)$$

where $\alpha_{\Delta t}$ is the Courant number in the CFL condition. The parameter $\prod_{ij}$ is computed as follows:

$$\prod_{ij} = \frac{\rho_j}{(\Delta t)^2 (\rho_i + \rho_j)} \left(\sqrt{d_{ij}^2 - \|\mathbf{r}_{ij\perp}^*\|^2} - \|\mathbf{r}_{ij\|}^*\| \right) \qquad (4.20)$$

where $\mathbf{r}^*_{ij\perp}$ is the normal vector of $\mathbf{r}^*_{ij}$ and $\mathbf{r}^*_{ij\parallel}$ is the parallel vector of $\mathbf{r}^*_{ij}$ with $\mathbf{r}^*_{ij} = \mathbf{r}^*_{ij\perp} + \mathbf{r}^*_{ij\parallel}$. More details regarding the dynamic stabilization can be found in Ref. [22].

4.5 Artificial viscosity

Artificial viscosity is a technique developed in the SPH framework to suppress the velocity fluctuations. It is effective to enhance the stability in particle methods. The original version of artificial viscosity was proposed by Monaghan [23]. The idea is rather simple yet effective: When two particles are approaching each other in the interaction radius, additional repulsive force is added between them. The model presented here is the one adjusted for the MPS method. Specifically the velocity is updated by the following equation after the calculation:

$$\mathbf{u}'_i = \mathbf{u}_i + \Delta\mathbf{u}_i \qquad (4.21)$$

where $\Delta\mathbf{u}_i$ is the velocity due to artificial viscosity. This term is calculated as follows:

$$\Delta\mathbf{u}_i = -\alpha\frac{d\Delta t}{n^0}\sum_{j\neq i}\prod_{ij}\frac{(\mathbf{r}_j - \mathbf{r}_i)}{r^2_{ij}}w_{ij} \qquad (4.22)$$

where $\prod_{ij}$ is computed as follows:

$$\prod_{ij} = \begin{cases} 0 & \mathbf{u}_{ij}\cdot\mathbf{r}_{ij} \geq 0 \\ -c_0 u_{ij} + 2u^2_{ij} & \mathbf{u}_{ij}\cdot\mathbf{r}_{ij} < 0 \end{cases} \qquad (4.23)$$

with

$$u_{ij} = \frac{h\mathbf{u}_{ij}\cdot\mathbf{r}_{ij}}{r^2_{ij} + 0.01h^2} \qquad (4.24)$$

In Eqs. (4.22)–(4.24), α is the parameter that determines the strength of artificial viscosity ($\alpha = 0.05$ is recommended), the characteristic length $h = 0.5r_e$, the sound speed is $c_0 = 10U_{max}$, and U_{max} is the maximum velocity in the considered flow field. This model is symmetric for a pair of particles. Thus it can guarantee the momentum conservation.

It is noted that the basic ideas of the collision model, the dynamic stabilization, and the artificial viscosity are essentially similar to each other.

For example, when two particles are approaching each other, additional repulsive forces will be supplemented to separate particles. The main differences lay in the conditions to invoke the repulsive forces and the method to determine the strength of repulsive forces. These terms can be applied to· all the free-surface, vicinity, and internal particles in Fig. 4.1. Nevertheless, these models can result in additional numerical dissipation to the simulations, which is the main drawback.

In brief the stabilization strategies in the MPS method are discussed. Because the particle stabilizing terms in the pressure gradient model (Section 4.1), the collision model (Section 4.3), the dynamic stabilization (Section 4.4), and the artificial viscosity (Section 4.5) usually produce additional numerical diffusion/dissipation, the particle shifting (Section 4.2) is a better technique for stability in this viewpoint. Meanwhile, there are two other advantages for particle shifting. First the effect of particle shifting can be considered by variable interpolation based on the Taylor series expansion [14]. Second the particle shifting techniques can effectively enhance the stability of the consistent high-order schemes. Therefore we recommend employing the particle shifting techniques as much as possible.

It must be noted that the free-surface-detection method can significantly influence the simulation stability [5]. Depending on the free-surface detection, two stable approaches are provided here. First, if the conventional free-surface-detection conditions are adopted, it is recommended to employ the conservative pressure gradient model for the free-surface and vicinity particles. Meanwhile the particle shifting techniques can be applied to all the particles and the consistent high-order schemes can be applied for the internal particles. This strategy was adopted in many studies [10,15−17,24−29]. Second, if one wants to apply the consistent high-order schemes to all the particles, the advanced free-surface-detection conditions based on the normalized coefficients of discretization models can be adopted [5]. In this situation the various particle shifting techniques including original PS, IPST, and OPS must be adopted for stability. Meanwhile, slight artificial viscosity can be adopted at free surfaces to suppress the fluctuations due to the free-surface detection. This strategy was adopted in the recent studies [5,9].

References

[1] J.W. Swegle, D.L. Hicks, S.W. Attaway, Smoothed particle hydrodynamics stability analysis, J. Comput. Phys. 116 (1995) 123−134.

[2] W. Dehnen, H. Aly, Improving convergence in smoothed particle hydrodynamics simulations without pairing instability, Mon. Not. R. Astron. Soc. 425 (2012) 1068–1082.

[3] S. Koshizuka, A. Nobe, Y. Oka, Numerical analysis of breaking waves using the moving particle semi-implicit method, Int. J. Numer. Methods Fluids 26 (1998) 751–769.

[4] A. Khayyer, H. Gotoh, Enhancement of stability and accuracy of the moving particle semi-implicit method, J. Comput. Phys. 230 (2011) 3093–3118.

[5] G. Duan, T. Matsunaga, S. Koshizuka, A. Yamaguchi, M. Sakai, New insights into error accumulation due to biased particle distribution in semi-implicit particle methods, Comput. Methods Appl. Mech. Eng. 388 (2022) 114291.

[6] S. Koshizuka, Y. Oka, Moving-particle semi-implicit method for fragmentation of incompressible fluid, Nucl. Sci. Eng. 123 (1996) 421–434.

[7] G. Duan, B. Chen, S. Koshizuka, H. Xiang, Stable multiphase moving particle semi-implicit method for incompressible interfacial flow, Comput. Methods in Appl. Mech. Eng. 381 (2017) 636–666.

[8] G. Duan, B. Chen, X. Zhang, Y. Wang, A multiphase MPS solver for modeling multi-fluid interaction with free surface and its application in oil spill, Comput. Methods Appl. Mech. Eng. 320 (2017) 133–161.

[9] G. Duan, M. Sakai, An enhanced semi-implicit particle method for simulating the flow of droplets with free surfaces, Comput. Methods Appl. Mech. Eng. (2022).

[10] G. Duan, S. Koshizuka, A. Yamaji, B. Chen, X. Li, T. Tamai, An accurate and stable multiphase moving particle semi-implicit method based on a corrective matrix for all particle interaction models, Int. J. Numer. Methods Eng. 115 (2018) 1287–1314.

[11] E. Toyota, A particle method with variable spatial resolution for incompressible flows, Proc. 19th Jpn. Soc. Fluid Mech. 9 (2005) 5–10.

[12] S.M. Jeong, J.W. Nam, S.C. Hwang, J.C. Park, M.H. Kim, Numerical prediction of oil amount leaked from a damaged tank using two-dimensional moving particle simulation method, Ocean Eng. 69 (2013) 70–78.

[13] K. Shibata, I. Masaie, M. Kondo, K. Murotani, S. Koshizuka, Improved pressure calculation for the moving particle semi-implicit method, Comput. Part. Mech. 2 (2015) 91–108.

[14] R. Xu, P. Stansby, D. Laurence, Accuracy and stability in incompressible SPH (ISPH) based on the projection method and a new approach, J. Comput. Phys. 228 (2009) 6703–6725.

[15] G. Duan, A. Yamaji, S. Koshizuka, A novel multiphase MPS algorithm for modeling crust formation by highly viscous fluid for simulating corium spreading, Nucl. Eng. Des. 343 (2019) 218–231.

[16] G. Duan, A. Yamaji, S. Koshizuka, B. Chen, The truncation and stabilization error in multiphase moving particle semi-implicit method based on corrective matrix: which is dominant? Comput. Fluids 190 (2019) 254–273.

[17] G. Duan, T. Matsunaga, A. Yamaji, S. Koshizuka, M. Sakai, Imposing accurate wall boundary conditions in corrective-matrix-based moving particle semi-implicit method for free surface flow, Int. J. Numer. Methods Fluids 93 (2021) 148–175.

[18] S.J. Lind, R. Xu, P.K. Stansby, B.D. Rogers, Incompressible smoothed particle hydrodynamics for free-surface flows: a generalised diffusion-based algorithm for stability and validations for impulsive flows and propagating waves, J. Comput. Phys. 231 (2012) 1499–1523.

[19] A. Khayyer, H. Gotoh, Y. Shimizu, Comparative study on accuracy and conservation properties of two particle regularization schemes and proposal of an optimized particle shifting scheme in ISPH context, J. Comput. Phys. 332 (2017) 236–256.

[20] P.P. Wang, Z.F. Meng, A.M. Zhang, F.R. Ming, P.N. Sun, Improved particle shifting technology and optimized free-surface detection method for free-surface flows in smoothed particle hydrodynamics, Comput. Methods Appl. Mech. Eng. 357 (2019) 112580.

[21] B.H. Lee, J.C. Park, M.H. Kim, S.C. Hwang, Step-by-step improvement of MPS method in simulating violent free-surface motions and impact-loads, Comput. Methods Appl. Mech. Eng. 200 (2011) 1113−1125.

[22] N. Tsuruta, A. Khayyer, H. Gotoh, A short note on dynamic stabilization of moving particle semi-implicit method, Comput. Fluids 82 (2013) 158−164.

[23] J.J. Monaghan, Simulating free surface flows with SPH, J. Comput. Phys. 110 (1994) 399−406.

[24] N. Takahashi, G. Duan, M. Furuya, A. Yamaji, Analysis of hemispherical vessel ablation failure involving natural convection by MPS method with corrective matrix, Int. J. Adv. Nucl. React. Des. Technol. 1 (2019) 19−29.

[25] Jubaidah, G. Duan, A. Yamaji, C. Journeau, L. Buffe, J.F. Haquet, Investigation on corium spreading over ceramic and concrete substrates in VULCANO VE-U7 experiment with moving particle semi-implicit method, Ann. Nucl. Energy 141 (2020) 107266.

[26] Jubaidah, Y. Umazume, N. Takahashi, X. Li, G. Duan, A. Yamaji, 2D MPS method analysis of ECOKATS-V1 spreading with crust fracture model, Nucl. Eng. Des. 379 (2021) 111251.

[27] R. Kawakami, X. Li, G. Duan, A. Yamaji, I. Sato, T. Suzuki, Improvement of solidification model and analysis of 3D channel blockage with MPS method, Front. Energy (2021) 1−13.

[28] N. Takahashi, G. Duan, A. Yamaji, X. Li, I. Sato, Development of MPS method and analytical approach for investigating RPV debris bed and lower head interaction in 1F Units-2 and 3, Nucl. Eng. Des. 379 (2021) 111244.

[29] G. Duan, A. Yamaji, M. Sakai, A multiphase MPS method coupling fluid−solid interaction/phase-change models with application to debris remelting in reactor lower plenum, Ann. Nucl. Energy 166 (2022) 108697.

CHAPTER FIVE

Boundary conditions

Boundary conditions play important roles in simulation since they are essential for solving practical problems. Although the moving particle semi-implicit (MPS) method can easily treat moving boundaries with large deformation, implementing boundary conditions could be a little bit problematic. This is because the effective domain is partially supported since the weight function is cut off near boundaries, as illustrated in Fig. 5.1. In this chapter the most commonly encountered physical boundaries, including the wall boundary, free-surface boundary, and inlet/outlet boundary, are introduced. Specifically, six widely used wall boundary models are explained in Section 5.1; Section 5.2 discusses three main approaches to deal with the free surface; in Section 5.3 the inflow and outflow boundaries are described.

5.1 Wall boundary

The solid wall boundary is the most fundamental one in computational fluid dynamics. In general the wall boundary plays two essential roles: enforcement of physical boundaries and compensation of the support domain. Various attempts have been made within the MPS method to model the wall boundary. As shown in Fig. 5.2, it can be mainly

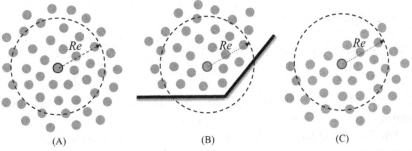

Figure 5.1 Support domain around a fluid particle. (A) Fully supported. (B) Wall boundary. (C) Free surface boundary.

Moving Particle Semi-implicit Method.
DOI: https://doi.org/10.1016/B978-0-443-13508-8.00005-6

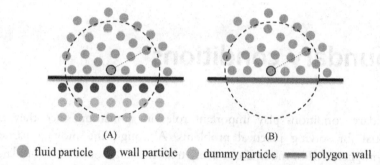

● fluid particle ● wall particle ● dummy particle ▬▬ polygon wall

Figure 5.2 Wall boundary models. (A) Particle wall. (B) Polygon wall.

classified into two types, that is, particle and polygon walls. The particle
wall model means that particles represent the wall boundary. In contrast
the polygon boundary is utilized in the polygon wall model.

5.1.1 Fixed wall (dummy) particle model

Fixed wall or dummy particles, proposed by Koshizuka and Oka [1,2], are
widely adopted. Usually, one-layer wall particles and several-layer dummy particles are placed along the computational domain to compensate the incomplete
weight support (see Fig. 5.2). The wall (dummy) particles are calculated similar
to fluid particles. Precisely the particle number density (PND) is calculated as

$$n_i = n_i^{\text{fluid}} + n_i^{\text{wall}} + n_i^{\text{dummy}} = \sum_{j \in \left(\Lambda_i^{\text{f}} \cup \Lambda_i^{\text{w}} \cup \Lambda_i^{\text{d}} \right)} w_{ij} \qquad (5.1)$$

The gradient operator:

$$\nabla \phi_i = \nabla \phi_i^{\text{fluid}} + \nabla \phi_i^{\text{wall}} + \nabla \phi_i^{\text{dummy}} = \frac{d}{n^0} \sum_{j \in \left(\Lambda_i^{\text{f}} \cup \Lambda_i^{\text{w}} \cup \Lambda_i^{\text{d}} \right)} \frac{\phi_{ij}}{r_{ij}^2} \mathbf{r}_{ij} w_{ij} \qquad (5.2)$$

The Laplacian operator:

$$\nabla^2 \phi_i = \nabla^2 \phi_i^{\text{fluid}} + \nabla^2 \phi_i^{\text{wall}} + \nabla^2 \phi_i^{\text{dummy}} = \frac{2d}{\lambda^0 n^0} \sum_{j \in \left(\Lambda_i^{\text{f}} \cup \Lambda_i^{\text{w}} \cup \Lambda_i^{\text{d}} \right)} \phi_{ij} w_{ij} \qquad (5.3)$$

Here, Λ_i^{f}, Λ_i^{w}, and Λ_i^{d} indicate the neighboring fluid, wall, and
dummy particles around particle i, respectively.

Dirichlet boundary condition is directly enforced by assigning the value
to the wall and dummy particles. For example, setting the wall (dummy)
particles at zero velocity can enforce the no–slip boundary. By contrast a

different treatment has been always adopted in the pressure calculation. More specifically the wall particle is involved in the pressure calculation, while the pressure of the dummy particle is assumed to be the same with the reference particle i:

$$P_j = P_i, \left(j \in \Lambda_{\text{dummy}}\right) \tag{5.4}$$

As a consequence the gradient and Laplacian of pressure can be rewritten as

$$\nabla P_i = \frac{d}{n^0} \left(\sum_{j \in \left(\Lambda_i^f \cup \Lambda_i^w\right)} \frac{P_j - \widehat{P}_i}{r_{ij}^2} \mathbf{r}_{ij} w_{ij} + \sum_{j \in \Lambda_i^d} \frac{P_i - \widehat{P}_i}{r_{ij}^2} \mathbf{r}_{ij} w_{ij} \right) \tag{5.5}$$

$$\nabla^2 P_i = \frac{2d}{\lambda^0 n^0} \sum_{j \in \left(\Lambda_i^f \cup \Lambda_i^w\right)} P_{ij} w_{ij} \tag{5.6}$$

where the minimum pressure $\widehat{P}_i$ is defined as

$$\widehat{P}_i = \min_{j \in (\Lambda_i \cup i)} P_j \tag{5.7}$$

Such treatment is straightforward and can provide reliable results for simple geometry. However, it is complicated and inefficient for complicated curved geometric shapes. Moreover the accuracy might be deteriorated in some cases. To be specific, the velocity no-slip condition is not accurately enforced because the zero-velocity wall particles are always not precisely located on the boundary. The Neumann boundary condition for pressure is also not accurately enforced because some assumptions are made regarding the pressure of dummy particles.

A similar and simple treatment, called ghost cell boundary model, has been developed by Zheng et al. [3]. In this model the wall (dummy) particles are replaced by the ghost cell. With such a treatment, it is relatively flexible in dealing with complex geometry and has been successfully coupled with the finite element method to solve the fluid—structure interaction (FSI) problems [4].

5.1.2 Mirror particle model

Mirror particle model is another wall boundary model. Unlike the fixed wall particles the boundary particles move along with the fluid particles. These moving boundary particles are called mirror particles because they can be regarded as mirror images of the (parent) fluid particles, as illustrated in Fig. 5.3. In general the discretization schemes for spatial

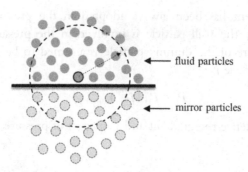

Figure 5.3 Mirror particles.

derivatives, that is, Eqs. (5.1–5.3), are the same as that in the fixed wall particle model. The physical quantities of mirror particles are determined based on the linear extrapolation from the fluid particles. For example the mirror velocity **u**′ for the no-slip boundary is determined simply by

$$\mathbf{u}' = 2\mathbf{u}_{\text{wall}} - \mathbf{u} \tag{5.8}$$

where **u** is the velocity on the parent fluid particle.

Different from the fixed wall particle model, there is no need to pre-define the wall particles in the mirror particle model. The mirror particles are determined according to the parent fluid particles and boundary geometry. Moreover, it is not necessary to define the computational variables for these mirror particles. This is because the quantities, for example, pressure and velocity, of mirror particles are based on linear extrapolation from the fluid particles. As a consequence the computational efficiency is improved. Although various advanced techniques have been developed [5–8], this treatment seems only effective in two-dimensional simulation. It remains challenging for the application to 3D complex geometry.

5.1.3 Fixed boundary particle model

The fixed boundary particles can be precisely set along the boundary to accurately enforce the boundary conditions, as illustrated in Fig. 5.4. This method is widely adopted in the least squares particle methods [9,10].

A Taylor series expansion of a function ϕ around $\mathbf{r}_i$ with a nearby particle $\mathbf{r}_j$ is written as

$$\phi_j = \phi_i + \sum_{m=1}^{p} \frac{1}{m!} \left(\mathbf{r}_{ij} \cdot \nabla\right)^m \phi|_{\mathbf{r}_i} + \mathrm{o}\left(\left|\mathbf{r}_{ij}\right|^{p+1}\right) \tag{5.9}$$

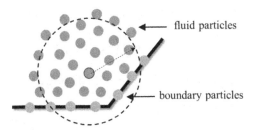

Figure 5.4 Fixed boundary particles.

If the higher-order terms are ignored, the following equation is obtained:

$$\phi_j - \phi_i = \sum_{m=1}^{p} \frac{r_s^m}{m!} \left(\frac{\mathbf{r}_{ij}}{r_s} \cdot \nabla \right)^m \phi|_{\mathbf{r}_i} = \mathbf{p}_{ij} \cdot \left(\mathbf{H}_{r_s}^{-1} \partial \phi|_{\mathbf{r}_i} \right) \qquad (5.10)$$

where $\phi_i = \phi(\mathbf{r}_i)$, $\phi_j = \phi(\mathbf{r}_j)$, $\mathbf{r}_{ij} = \mathbf{r}_j - \mathbf{r}_i$, r_s denotes a scaling parameter, p is a positive integer, which indicates the order of approximation, $\mathbf{p}_{ij}$ denotes a polynomial basis vector, $\mathbf{H}_{r_s}$ represents a scaling matrix, and ∂ indicates a differential operator vector.

For the second-order scheme ($p = 2$), these functions are written as

$$\mathbf{p}_{ij} = \mathbf{p} \left(\frac{\mathbf{r}_{ij}}{r_s} \right) = \left[\frac{x_{ij}}{r_s}, \frac{y_{ij}}{r_s}, \frac{x_{ij}^2}{r_s^2}, \frac{x_{ij}y_{ij}}{r_s^2}, \frac{y_{ij}^2}{r_s^2} \right]^T \qquad (5.11)$$

$$\mathbf{H}_{r_s} = \mathrm{diag} \left(\frac{1}{r_s}, \frac{1}{r_s}, \frac{2}{r_s^2}, \frac{1}{r_s^2}, \frac{2}{r_s^2} \right) \qquad (5.12)$$

$$\partial = \left[\frac{\partial}{\partial x}, \frac{\partial}{\partial y}, \frac{\partial^2}{\partial x^2}, \frac{\partial^2}{\partial x \partial y}, \frac{\partial^2}{\partial y^2} \right]^T \qquad (5.13)$$

If the gradient operator is applied on both sides in Eq. (5.9), the following equation is obtained:

$$\nabla \phi|_{\mathbf{r}=\mathbf{r}_j} = \nabla \phi|_{\mathbf{r}=\mathbf{r}_i} + \sum_{m=1}^{p-1} \frac{1}{m!} \left(\mathbf{r}_{ij} \cdot \nabla \right)^m \nabla \phi|_{\mathbf{r}=\mathbf{r}_i} + \mathrm{o}\left(|\mathbf{r}_{ij}|^p \right)$$

$$= \sum_{m=0}^{p-1} \frac{1}{m!} \left(\mathbf{r}_{ij} \cdot \nabla \right)^m \nabla \phi|_{\mathbf{r}=\mathbf{r}_i} + \mathrm{o}\left(|\mathbf{r}_{ij}|^p \right) \qquad (5.14)$$

Multiplying both sides by the unit normal vector $\mathbf{n}_j$, Eq. (5.14) is transformed to

$$\frac{\partial \phi}{\partial \mathbf{n}}|_{\mathbf{r}_j} = \mathbf{n}_j \sum_{m=0}^{p-1} \frac{r_s^m}{m!} \left(\frac{\mathbf{r}_{ij}}{r_s} \cdot \nabla \right)^m \nabla \phi|_{\mathbf{r}_i} + \mathrm{o}\left(|\mathbf{r}_{ij}|^p \right) \qquad (5.15)$$

By ignoring the higher-order terms, Eq. (5.15) can be rewritten as

$$r_s \frac{\partial \phi}{\partial \mathbf{n}}\Big|_{\mathbf{r}=\mathbf{r}_j} = \mathbf{q}_{ij} \cdot \left(\mathbf{H}_{r_s}^{-1} \partial \phi |_{\mathbf{r}_i} \right) \tag{5.16}$$

where $\mathbf{q}_{ij} = \frac{\partial}{\partial \mathbf{n}} \left(r_s \mathbf{p}_{ij} \right)$ indicates a modified polynomial basis vector. Specifically, $\mathbf{q}_{ij}$ for $p = 2$ is given as

$$\mathbf{q}_{ij} = \left[n_x, n_y, 2n_x \frac{x_{ij}}{r_s}, n_x \frac{y_{ij}}{r_s} + n_y \frac{x_{ij}}{r_s}, 2n_y \frac{y_{ij}}{r_s} \right]^T \tag{5.17}$$

The Dirichlet boundary condition can be treated simply based on the least squares scheme by substituting the generated boundary particles with the prescribed boundary values. As for the Neumann boundary condition it can be constrainted into the boundary particles using the constraint scheme [11] or the Lagrange multiplier [9].

The spatial derivatives for the boundary particles using the constraint scheme are given as

$$\partial \phi_{\mathbf{r}=\mathbf{r}_i} = \mathbf{H}_{r_s} (\mathbf{M}_i + \mathbf{N}_i)^{-1} (\mathbf{b}_i + \mathbf{c}_i) \tag{5.18}$$

with

$$\mathbf{M}_i = \sum_{j \neq i} w_{ij} \mathbf{p}_{ij} \otimes \mathbf{p}_{ij} \tag{5.19}$$

$$\mathbf{N}_i = \sum_{j=i} w_0 \mathbf{q}_{ij} \otimes \mathbf{q}_{ij} \tag{5.20}$$

$$\mathbf{b}_i = \sum_{j \neq i} w_{ij} \mathbf{p}_{ij} \left(\phi_j - \phi_i \right) \tag{5.21}$$

$$\mathbf{c}_i = \sum_{j=i} w_0 \mathbf{q}_{ij} r_s \frac{\partial \phi}{\partial \mathbf{n}} \tag{5.22}$$

where w_0 determines the relative importance of the constraint, which is usually set to be w_{ij} (0). Compared with the constraint scheme the boundary condition is strictly satisfied without relying on any tuning parameter using the Lagrange multiplier. In this situation the spatial derivatives for the boundary particles are given as

$$\check{\partial} \phi_{\mathbf{r}=\mathbf{r}_i} = \check{\mathbf{H}}_{r_s} \check{\mathbf{M}}_i^{-1} \check{\mathbf{b}}_i \tag{5.23}$$

with

$$\check{\mathbf{M}}_i = \begin{bmatrix} \mathbf{M}_i \mathbf{q}_{ij} \\ \mathbf{q}_{ij}^T 0 \end{bmatrix} \tag{5.24}$$

$$\breve{\partial} = [\partial | \lambda]^T \qquad (5.25)$$

$$\breve{\mathbf{H}}_{r_s} = \mathrm{diag}\left(\mathbf{H}_{r_s} | \lambda_{r_s}\right) \qquad (5.26)$$

$$\breve{\mathbf{b}}_i = \left[\mathbf{b}_i | r_s \frac{\partial \phi}{\partial \mathbf{n}}\right]^T \qquad (5.27)$$

where λ indicates the Lagrange multiplier and λ_{r_s} indicates the scaling parameter, both of which are not involved in calculating the spatial derivatives. Note that for both the constraint scheme and the Lagrange multiplier, only the boundary particle itself is considered for the modified polynomial basis vector $\mathbf{q}_{ij}$, namely, $x_{ij} = 0$, $y_{ij} = 0$, as

$$\mathbf{q}_{ij} = \left[n_x, n_y, 0, 0, 0\right]^T \qquad (5.28)$$

This kind of treatment is straightforward. However, the boundary particles should be carefully predefined for complex geometry. Otherwise, some potential numerical instabilities, for example, due to not satisfying the positivity conditions [12,13], might be triggered.

5.1.4 Distance-based polygon model

A distance-based polygon wall boundary model has been developed by Harada et al. [14] with the aim to efficiently and stably deal with complex geometry. In this model the wall boundary is expressed as a polygon rather than particles. The interaction between a fluid particle and wall boundary is estimated based on the closest distance. Compared with the above-mentioned particle-based models, it is much easier and more flexible to handle the complex wall boundary geometry with the help of computer-aided design (CAD) tools.

The wall contribution in the PND n_i^{boundary} is approximated by the so-called wall weight function $Z(|\mathbf{r}_{iw}|)$. The wall weight function is calculated by assuming the fixed wall particles placed outside a flat boundary, as shown in Fig. 5.5. Note that the wall weight function can be prestored to save computational costs.

$$n_i^{\text{boundary}} = \sum\nolimits_{j \in \Lambda_i^{\text{boundary}}} w_{ij} \approx Z(|\mathbf{r}_{iw}|) \qquad (5.29)$$

Meanwhile, to prevent the fluid particle from penetrating the wall boundary a repulsive contact force is considered to push the fluid particle

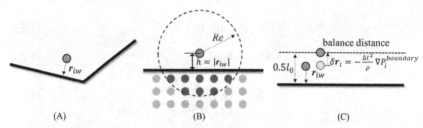

Figure 5.5 Schematic illustration of distance-based polygon model. (A) Closed point on wall. (B) Fixed wall particle. (C) Position correction.

back to a certain balance distance from the wall (usually is $0.5l_0$):

$$\delta \mathbf{r}_i = -\frac{\Delta t^2}{\rho} \nabla P_i^{\text{boundary}} \tag{5.30}$$

By doing this the gradient model and Laplacian model between fluid particles and wall boundaries are modified in relation to the distance function as

$$\nabla P_i^{\text{boundary}} = \begin{cases} \dfrac{\rho}{\Delta t^2} \dfrac{\mathbf{r}_{iw}}{|\mathbf{r}_{iw}|}(0.5l_0 - |\mathbf{r}_{iw}|) & |\mathbf{r}_{iw}| < 0.5l_0 \\ 0 & |\mathbf{r}_{iw}| \geq 0.5l_0 \end{cases} \tag{5.31}$$

$$\nabla^2 P_i^{\text{boundary}} = \begin{cases} \dfrac{2\rho}{\lambda^0} \dfrac{|\mathbf{r}_{iw}|}{\Delta t^2}(0.5l_0 - |\mathbf{r}_{iw}|) & |\mathbf{r}_{iw}| < 0.5l_0 \\ 0 & |\mathbf{r}_{iw}| \geq 0.5l_0 \end{cases} \tag{5.32}$$

$$\nabla^2 \mathbf{u}_i^{\text{boundary}} = \frac{2d}{\lambda^0 n^0}(\mathbf{u}_{\text{boundary}} - \mathbf{u}_i)Z(|\mathbf{r}_{iw}|) \tag{5.33}$$

The distance-based polygon wall boundary model has been applied to a wide range of engineering problems, despite that some unphysical behaviors are observed in the presence of corners. Although some advanced techniques have been developed, such as the work by Zhang et al. [15–17], the accuracy remains an issue.

5.1.5 Integral-based polygon model

In order to accurately deal with complex geometry an integral-based polygon wall boundary model has been developed by Matsunaga et al. [18,19]. As shown in Fig. 5.6 the main idea to consider the wall contribution is transforming the volume integration into surface integration.

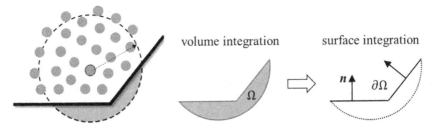

Figure 5.6 Transformation of volume integration into surface integration using the divergence theorem.

Based on the divergence theorem the wall contribution for the discretization schemes, including the PND, pressure gradient, Laplacian of velocity, and Laplacian of pressure, can be rewritten as

$$n_i^{\text{boundary}} = \frac{1}{C_V}\sum\nolimits_k (-h_{ik})\int_{\partial\Omega_i^k} f_m(|\mathbf{x}-\mathbf{r}_i|)dS \tag{5.34}$$

$$\nabla P_i^{\text{boundary}} = \frac{m}{n^0 C_V}\sum\nolimits_k \nabla P|_{\text{wall}}^k (-h_{ik})\int_{\partial\Omega_i^k} \frac{(\mathbf{x}-\mathbf{r}_i)\otimes(\mathbf{x}-\mathbf{r}_i)}{|\mathbf{x}-\mathbf{r}_i|^2} f_m(|\mathbf{x}-\mathbf{r}_i|)dS$$

$$\tag{5.35}$$

$$\nabla^2 P_i^{\text{boundary}} = \frac{2m}{\lambda^0 n^0 C_V}\sum\nolimits_k \nabla P|_{\text{wall}}^k (-h_{ik})\int_{\partial\Omega_i^k} f_{m+1}(|\mathbf{x}-\mathbf{r}_i|)dS \tag{5.36}$$

$$\nabla\cdot\mathbf{u}_i^{\text{boundary}} = \frac{m}{n^0 C_V}\sum\nolimits_k (\mathbf{u}_{\text{wall}}^k - \mathbf{u}_i)(-h_{ik})\int_{\partial\Omega_i^k} \frac{(\mathbf{x}-\mathbf{r}_i)}{|\mathbf{x}-\mathbf{r}_i|^2} f_m(|\mathbf{x}-\mathbf{r}_i|)dS$$

$$\tag{5.37}$$

$$\nabla^2\mathbf{u}_i^{\text{boundary}} = \frac{2m}{\lambda^0 n^0 C_V}\sum\nolimits_k (\mathbf{u}_{\text{wall}}^k - \mathbf{u}_i)(-h_{ik})\int_{\partial\Omega_i^k} f_{m+1}(|\mathbf{x}-\mathbf{r}_i|)dS \tag{5.38}$$

with

$$C_V = \frac{1}{Z(h)}\int_{\Omega(h)} w(|\mathbf{x}|)dV \tag{5.39}$$

$$f_m = \frac{1}{r^m}\int r^m w(r)dr \tag{5.40}$$

where C_V is the scaling coefficient, $Z(h)$ is the wall weight function, given in Eq. (5.29). The boundary integrals in Eqs. (5.34−5.40) can be solved analytically [19] or using the Gaussian quadrature [18] for a given weight function w.

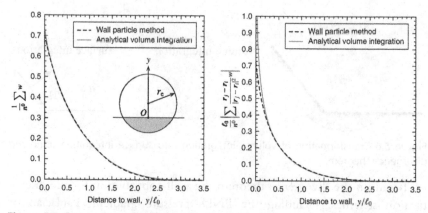

Figure 5.7 Comparison of wall contributions with different boundary treatments for a wall corner with a concave angle [19].

Fig. 5.7 compares wall contributions with the fixed wall particle method, distance-based polygon method, and integral-based polygon method for a wall corner with a concave angle. Compared with the distance-based polygon method, more accurate results are obtained in the integral-based polygon method. As a result, this method can provide accurate and robust results for application to arbitrary geometry in 2D. However, the extension to 3D might not be easy.

5.1.6 Virtual particle-based polygon model

The virtual wall particle-based polygon model has been developed by Matsunaga et al. [20] to deal with arbitrary geometry in 2D and 3D. As illustrated in Fig. 5.8 the virtual boundary particles are assumed to be the closest points, dynamically and locally generated on the boundaries. No computational variables are defined for the virtual boundary particles. The schemes only consider the contributions from these boundary particles to nearby fluid particles but not vice versa. With this treatment, complicated curved geometries can be easily and accurately handled.

By considering all neighboring particles and nodes as j ($j \neq i$) for Eq. (5.10) and Eq. (5.16) the objective function J for a weighted least squares problem is defined as

$$J(\mathbf{X}_i) = \sum_{j \neq i, j \in \Lambda_F} w_{ij} \left(\mathbf{p}_{ij} \cdot \mathbf{X}_i - \phi_j + \phi_i \right)^2 + \sum_{j \in \Lambda_B} w_{ij} \left(\mathbf{q}_{ij} \cdot \mathbf{X}_i - r_s \frac{\partial \phi}{\partial \mathbf{n}} |_{\mathbf{r} = \mathbf{r}_j} \right)^2$$

$$(5.41)$$

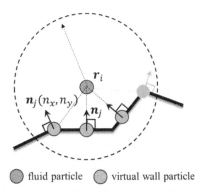

● fluid particle   ○ virtual wall particle

Figure 5.8 Virtual wall particle-based polygon model [20].

with

$$\mathbf{X}_i = \mathbf{H}_{r_s}^{-1}\partial\phi|_{\mathbf{r}_i} \tag{5.42}$$

where Λ_F and Λ_B indicate fluid particles and virtual boundary particles, respectively.

Minimization of J leads to the normal equations:

$$(\mathbf{M}_i + \mathbf{N}_i)\mathbf{X}_i = (\mathbf{b}_i + \mathbf{c}_i) \tag{5.43}$$

with

$$\mathbf{M}_i = \sum_{j\neq i, j\in\Lambda_F} w_{ij}\mathbf{p}_{ij}\otimes\mathbf{p}_{ij} \tag{5.44}$$

$$\mathbf{N}_i = \sum_{j\in\Lambda_B} w_{ij}\mathbf{q}_{ij}\otimes\mathbf{q}_{ij} \tag{5.45}$$

$$\mathbf{b}_i = \sum_{j\neq i, j\in\Lambda_F} w_{ij}\mathbf{p}_{ij}\left(\phi_j - \phi_i\right) \tag{5.46}$$

$$\mathbf{c}_i = \sum_{j\in\Lambda_B} w_{ij}\mathbf{q}_{ij}r_s\frac{\partial\phi}{\partial\mathbf{n}}|_{\mathbf{r}=\mathbf{r}_j} \tag{5.47}$$

For the least squares problem the weight is only used to weigh the squared error at given points in constructing the objective function. Consequently the weight corresponding to the wall region does not need to match the total weight over the solid domain precisely. A single wall particle can represent the influence from the wall region with the same weight function as employed for fluid particles. Compared with the previous fixed boundary particle model, there is no need to explicitly position boundary particles in advance. Numerical stability could be improved further. Specifically, there is no need to solve the linear system defined for

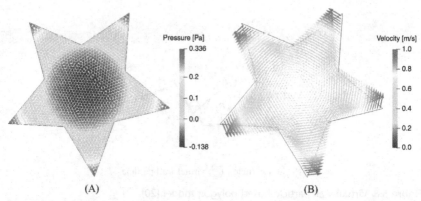

Figure 5.9 Pressure and velocity in a rigid rotation problem [20]. (A) Pressure. (B) Velocity.

the boundary particles. In addition the spatial derivatives can be calculated explicitly.

Fig. 5.9 shows a complex rigid rotation problem with complex geometry [20]. Excellent rigid rotational motion has been obtained for the internal flow. As addressed by Duan et al. [21], this kind of treatment can be also applied to the corrective matrix-based MPS method.

A similar boundary treatment, called copy particle technique, has been proposed by Tanaka et al. [22]. In their treatment the boundary particles are dynamically generated by projecting nearby fluid particles onto the wall boundaries. With this, complex geometries can be handled easily as well.

5.2 Free surface boundary

Free-surface flows are common in daily life and engineering applications. In the free-surface flow the gas pressure is assumed constant, and the viscous stress is assumed negligible. This idealization significantly reduces the complexity of the problem. Nevertheless, the treatment of free-surface boundary would influence accuracy and stability.

5.2.1 Free-surface particle detection

In the MPS method the most common way to impose the free-surface boundary is by detecting the free-surface particles. The pressure of the detected free-surface particles is set to be the background pressure. In other words, the Dirichlet boundary conditions are given to the free-surface particles when solving the pressure Poisson equation (PPE). Due to the incomplete support domain in the vicinity of the free surface, free-surface particles can be detected by PND decrease, as described in Section 2.3.1. This kind of treatment is straightforward and efficient. However, some internal particles may be misjudged as free-surface particles. Another often adopted detection criterion proposed by Tanaka and Masunaga [23] is

$$N_i < \beta' N^0 \tag{5.48}$$

where N_i is the number of neighboring particles, N^0 is the value in the initial particle arrangement, and β' is a coefficient.

Meanwhile, some accurate detection methods based on geometrical information can be found in Refs. [11,24−26]. For example, the surface normal of particle i can be computed from

$$\mathbf{n}_i = \frac{\mathbf{N}_i}{\|\mathbf{N}_i\|}, \mathbf{N}_i = \frac{1}{n_i} \sum_{j \neq i} \left[\frac{\mathbf{r}_j - \mathbf{r}_i}{r_{ij}} w_{ij} \right] \tag{5.49}$$

As for the free-surface detection, the following conditions are adopted:

$$\begin{cases} \text{if} \left(n_i < 0.97 n_0 \text{ or } N_i < 0.85 N_0 \text{ or } \|\mathbf{N}_i\| > 0.05 \right) & \text{parachute region detection} \\ \text{else} & \text{internal particle} \end{cases}$$

$$\tag{5.50}$$

where n_i is the PND, $\mathbf{N}_i$ is the surface normal vector, and $\mathbf{n}_i$ is the unit normal vector. As illustrated in Fig. 5.10 the parachute region detection tests whether a neighboring particle exists in the parachute region based

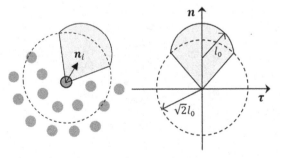

Figure 5.10 Parachute-shaped scan region.

on the following conditions [11,25]:

$$\begin{cases} \|\mathbf{r}_j - \mathbf{r}_i\| \geq \sqrt{2}l_0 \\ \|\mathbf{r}_i + l_0\mathbf{n}_i - \mathbf{r}_j\| < l_0 \end{cases} \tag{5.51}$$

or

$$\begin{cases} \|\mathbf{r}_j - \mathbf{r}_i\| < \sqrt{2}l_0 \\ \dfrac{\mathbf{r}_j - \mathbf{r}_i}{\|\mathbf{r}_j - \mathbf{r}_i\|} \cdot \mathbf{n}_i > \dfrac{1}{\sqrt{2}} \end{cases} \tag{5.52}$$

If Eq. (5.51) or Eq. (5.52) is satisfied, particle i is detected as an internal particle; otherwise, particle i is a free-surface particle.

Based on the detected free-surface particles, Shibata et al. [27] defined A-type and B-type free-surface particles, as shown in Fig. 5.11. A-type free-surface particles are the outermost layer of the fluid particles. B-type particles are defined as the ones in the vicinity of A-type but not A-type themselves. To avoid numerical instability, negative pressure is set to zero for the (A-type and B-type) free-surface particles.

If the corrective matrix (or the LS scheme) is adopted, careful attention should be paid to the free-surface boundary. Otherwise, potential numerical instability would occur due to the incomplete or biased neighbor support. Duan et al. [28] suggested to use the conservation pressure gradient scheme for free-surface particles (A-type and B-type). That is to say, the corrective matrix is imposed only for the internal fluid particles.

$$\nabla P_i = \begin{cases} \dfrac{d}{n^0} \sum_{j \neq i} \dfrac{P_j + P_i}{r_{ij}^2} \mathbf{r}_{ij} w_{ij} & \text{free surface particles} \\ \dfrac{d}{n^0} \sum_{j \neq i} \left(\dfrac{P_j - P_i}{r_{ij}} \mathbf{C}_i \dfrac{\mathbf{r}_{ij}}{r_{ij}} w_{ij} \right) & \text{internal fluid particles} \end{cases} \tag{5.53}$$

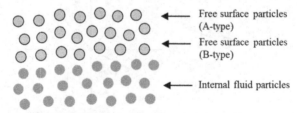

Figure 5.11 A-type and B-type free-surface particles.

$\mathbf{C}_i^{-1}$ represents the corrective matrix [28], and the formulation for the first-order scheme is given as

$$
\mathbf{C}_i^{-1} = \frac{d}{n^0}
\begin{bmatrix}
\sum_{j\neq i} \dfrac{x_{ij}^2}{r_{ij}^2} w_{ij} & \sum_{j\neq i} \dfrac{x_{ij} y_{ij}}{r_{ij}^2} w_{ij} \\
\sum_{j\neq i} \dfrac{y_{ij} x_{ij}}{r_{ij}^2} w_{ij} & \sum_{j\neq i} \dfrac{y_{ij}^2}{r_{ij}^2} w_{ij}
\end{bmatrix}
\tag{5.54}
$$

Recently a novel basis-based free-surface detection method has been proposed for the MPS with a corrective matrix by Duan et al. [29]. Error analysis is adopted to improve both stability and accuracy in their method. In other words the particles with high error-accumulation risk are detected as free-surface particles to guarantee stability. The recommended conditions for free-surface detection are given as follows:

$$
\begin{cases}
\text{if} \left(\|\mathbf{b}_i\| > 2.0 \text{ or } R_i > 1.5 \text{ or } c_i < 3.0 \text{ or } L_i < 0.25 \right) & \text{free surface particle} \\
\text{else if} (N_i < 0.35 N_0) & \text{parachute region} \\
\text{else} & \text{detection internal particle}
\end{cases}
\tag{5.55}
$$

$$
\mathbf{b}_i = \sum_{j \in N_D} \left[\frac{r_e}{r_{ij}} \frac{w_{ij}}{n_0} \left(\begin{bmatrix} \mathbf{M}_1 \\ \mathbf{M}_2 \end{bmatrix} \mathbf{p} \right) \right]
\tag{5.56}
$$

$$
c_i = \sum_{j \in N_D} \left(r_e l_0 \Gamma_{ij} \right)
\tag{5.57}
$$

$$
R_i = \frac{\|\mathbf{b}_i\|^2}{c_i}
\tag{5.58}
$$

$$
L_i = \frac{\sum_{j \in N_D} \Gamma_{ij}}{\sum_{j \in N_D} |\Gamma_{ij}|}
\tag{5.59}
$$

$$
\Gamma_{ij} = 2 \frac{w_{ij}}{n_0} \frac{[\mathbf{M}_3 + \mathbf{M}_4] \mathbf{p}}{r_{ij} l_0}
\tag{5.60}
$$

where parachute-region detection is the same as in Eqs. (5.51 and 5.52), $\mathbf{p}$ denotes the polynomial basis vector, and $\mathbf{M_k}$ is the k-th row of the inversed matrix. The restriction on R_i is used to suppress the fast error accumulation, while the restriction on $\mathbf{b}_i$ is used to suppress the significant bias error due to sudden pressure fluctuations when free surfaces impact each other. The restrictions on c_i and L_i are used to suppress unphysical negative diffusion and guarantee a well-conditioned PPE system. The "parachute-region detection" is used to suppress the fluctuations at free surfaces due to the reduced discretization accuracy when the neighboring-particle number significantly decreases. Fig. 5.12 shows dam break simulation results with the existing and bias-based free-surface detection methods. Unphysical large velocities take place suddenly and frequently without using the bias-based free-surface detection method. On the other hand, even though fewer free-surface particles are detected, more stable results are obtained with the bias-based free-surface detection method. It is worth mentioning that with this bias-based free-surface detection method, the corrective matrix can be applied to all the fluid particles, including free surface and internal particles.

5.2.2 Virtual free-surface particle

Instead of detecting the free-surface particles the so-called conceptual particle or virtual particle methods have been proposed to enhance the accuracy of the free-surface boundary in Refs. [27,30,31]. In such methods the conceptual or virtual particles rather than fluid particles provide the free-surface boundary condition, as shown in Fig. 5.13. Considering the effect of virtual particles a modified PND is defined as

$$n'_i = \max[n_i, \tilde{n}_i] \tag{5.61}$$

$$\tilde{n}_i = n_0 + \sum_{j \in \Lambda_i} \max\left(w_{ij} - w(l_0), 0\right) \tag{5.62}$$

The modified PPE for a free-surface particle is given as

$$\frac{2d}{\lambda^0 n^0} \sum_{j \in \Lambda_i^f \cup \Lambda_i^c} \left(P_j - P_i\right) w_{ij} = (1 - \gamma)\nabla \cdot \mathbf{u}_i^* + \gamma \frac{\rho}{\Delta t^2} \frac{n^0 - n'_i}{n^0} \tag{5.63}$$

A constant pressure, denoted by P_{gas}, is assumed for the virtual particles. Eq. (5.63) can be rewritten as

$$\frac{2d}{\lambda^0 n^0} \sum_{j \in \Lambda_i^f} \left(P_j - P_i\right) w_{ij} - \frac{2d}{\lambda^0 n^0} \left(n'_i - n_i\right) P_i = (1 - \gamma)\nabla \cdot \mathbf{u}_i^* + \gamma \frac{\rho}{\Delta t^2} \frac{n^0 - n'_i}{n^0}$$

$$- \frac{2d}{\lambda^0 n^0} \left(n'_i - n_i\right) P_{\text{gas}}$$

$$\tag{5.64}$$

As suggested by Chen et al. [30] the conservative pressure gradient scheme proposed by Toyota et al. [32], rather than the conventional

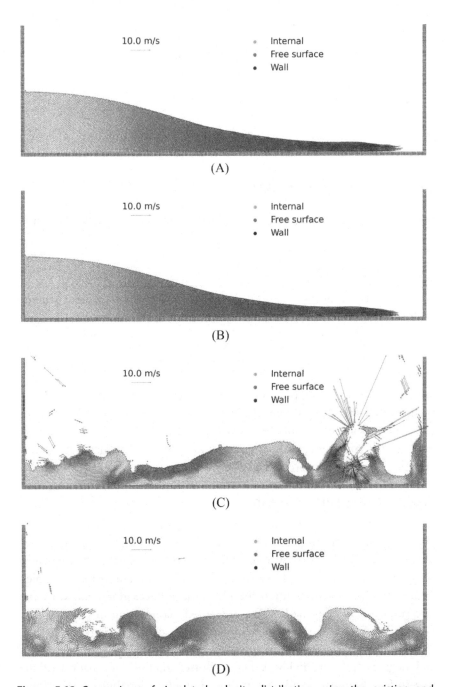

Figure 5.12 Comparison of simulated velocity distribution using the existing and bias-based free-surface detection methods [29]. (A) Existing conditions at $t = 0.57$ s. (B) Proposed conditions at $t = 0.57$ s. (C) Existing conditions at $t = 2.68$ s. (D) Proposed conditions at $t = 2.68$ s.

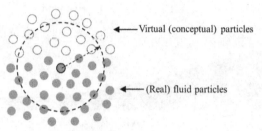

Figure 5.13 Virtual (conceptual) particles.

pressure gradient scheme, is more suitable for preventing particle clustering:

$$\nabla P_i = \frac{d}{n^0} \sum_{j \in \Lambda_i} \frac{P_j + P_i}{r_{ij}^2} \mathbf{r}_{ij} w_{ij} \tag{5.65}$$

Meanwhile, Shibata et al. [27] defined a new pressure gradient scheme with the contributions of virtual particles as

$$\nabla P_i = \frac{d}{n^0} \sum_{j \in \Lambda_i} \frac{P_j}{r_{ij}^2} \mathbf{r}_{ij} w_{ij} \tag{5.66}$$

which also helps to suppress instabilities at free surfaces. A similar treatment, called the space potential particles, has been proposed by Tsuruta et al. [31]. Different from the virtual particles the specific locations are given to the space potential particles, which provide the Dirichlet boundary conditions for pressure calculations.

5.2.3 Moving surface mesh

As opposed to the free-surface detection a moving surface mesh (see Fig. 5.14) has been proposed by Matsunaga et al. [33] to explicitly represent the deformable free surface. The accuracy is remarkably improved since the surface nodes, defined on the surface mesh, are strongly coupled with fluid particles. In addition the free-surface fluctuations caused by the free-surface detection method are prevented via inserting or deleting surface nodes. Fig. 5.15 shows the simulation of square patch rotation [34]. This simulation is not easy because the rotation leads to centrifugal forces and negative pressure fields. What is worse, intense deformation of the free-surface boundary could easily cause numerical instabilities. By adopting the moving surface mesh, highly deformed free surfaces were stably simulated. The four corner nodes satisfactorily followed the theoretical trajectory lines. Another advantage of this method is that the surface

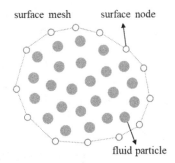

Figure 5.14 Schematic illustration of the moving surface mesh, consisting of surface nodes.

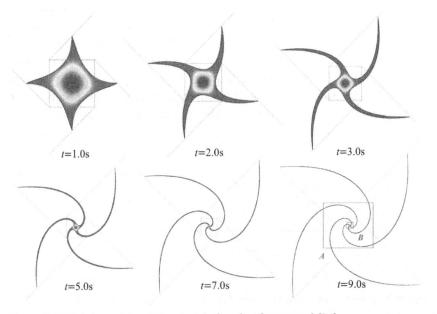

Figure 5.15 Square patch rotation test (colored with pressure) [34].

tension can be directly estimated based on the geometry of the surface mesh, which is introduced in Section 6.3.

5.3 Inflow and outflow boundaries

Inflow and outflow boundaries widely exist in engineering, such as pipe flow, fluid machinery, and heat exchangers. They can be easily implemented in the Euler-based numerical method. However, special treatments are necessary in the particle method because particles should flow in and out at a certain Eulerian position. Shakibaeinia and Jin [35] proposed a particle recycling strategy for inflow and outflow boundaries, as shown in Fig. 5.16. An extra type of particle, for example, storage particles, is used for particle cycling, which has no physical properties. When fluid particles flow out of the computation domain, they change to storage particles and the relevant physical properties are removed. Conversely, storage particles are changed to fluid particles and the physical properties are imposed on at the inflow boundary. Fig. 5.17 shows the conditions at inflow and outflow boundaries. Some layers of fixed-positioned ghost particles are defined at the inflow and outflow boundaries to compensate the density deficiency of fluid particles. For the known velocity inflow boundary the ghost particles move inward with the same boundary velocity. When the displacement distance of ghost particles is equal to the particle average distance the ghost particles go back to their initial positions and new fluid particles are added to the gap between the first layer of ghost particles and fluid particles. The physical information carried by

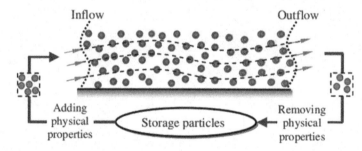

Figure 5.16 Particle recycling strategy [35].

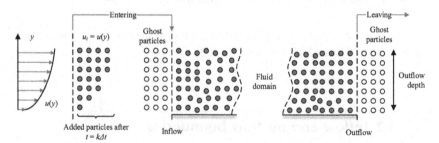

Figure 5.17 Inflow and outflow boundary conditions [35]. (A) Known velocity inflow boundary condition and (B) known pressure boundary condition.

these new generated particles is prescribed according to boundary conditions or extrapolated from the fluid particles in the main domain. For the known pressure boundary conditions the particles are added to inflow to compensate for particle deficiency resulting from the movement of particles to the main domain. At the outflow boundary the fluid particles that approach the first layer of ghost particles are removed. The boundary pressure that has been prescribed to the ghost particles is transferred to the fluid particles by repulsive forces.

Shibata et al. [36] and Duan et al. [37] improved Shakibaeinia's inflow boundary by setting a layer of pressure particles between the ghost and fluid particles. The pressure particles are involved in the pressure calculation as fluid particles. With this treatment the pressure oscillation caused by particle type change at the inflow boundary can be reduced. Tanaka et al. [22] proposed to set pressure boundaries by dynamically generating and removing particles at the boundary panel, and the Dirichlet boundary condition is imposed onto the pressure boundary particles.

Given the problem of fluid flowing around an object, the unphysical voids may form behind the object in simulation with the standard MPS method. Fig. 5.18 shows the unnatural cavities in Karman vortex flow [36]. It is because the outflow particles move out at their own velocities, resulting in the nonconservation of fluid particles in the main domain. Shibata et al. [36] optimized outflow boundaries by imposing a fixed velocity to the fluid particles that are close to the boundary line. The outflow velocity is calculated from the velocities of upstream particles, and a main criterion is to satisfy mass conservation of fluids in the channel.

$$\mathbf{u}_i^* \cdot \mathbf{n} = (\mathbf{u}_i^{*\text{upstream}} \cdot \mathbf{n} - \Delta u) \qquad (5.67)$$

$$\Delta u = \frac{(N^0 - N^*)l_0^3}{A} \frac{1}{\alpha \Delta t} \qquad (5.68)$$

where vector $\mathbf{u}_i^*$ is the temporary velocity of particle i located in the fixed velocity region on the outflow boundary, vector $\mathbf{n}$ is the direction of the

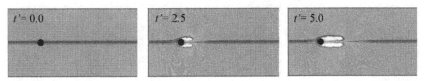

Figure 5.18 Unnatural cavity exists in the Karman vortex flow (the fluid flows from left to right with $Re = 100$) [36].

main stream, N^0 and N^* are the initial and current particle numbers in the entire domain, parameter A is the cross-sectional area of the outflow boundary, parameter α is a relaxation coefficient, parameter Δu is the correction term for conserving the amount of fluid in the simulation domain, and $\mathbf{u}_i^{*\text{upstream}}$ is the average velocity of an upstream. The $\mathbf{u}_i^{*\text{upstream}}$ can be calculated by the average of velocities of neighboring particles. Regarding how long is the distance to the outflow line to be defined as the upstream, it is usually set as 2 times the effective radius. In order to prevent harmful disturbance to the upstream, the upstream velocity distribution is approximately given to the outflow particles. When the value of $\mathbf{u}_i^* \cdot \mathbf{n}$ is smaller than zero, it will be set as zero to avoid backflow. Fig. 5.19 shows the Karman vortices simulated with the improved outflow boundary, in which the unnatural cavities are successfully avoided.

Shibata et al. [38] developed an inflow—outflow boundary for overlapping particle technique of multiresolution scheme. One-layer cells are defined to generate inflow particles, which are square and cubic shapes, respectively, for two and three dimensions. The length of each cell edge L_{Cell} is equal to the average particle spacing l_0. Many candidate position points are distributed in one inflow cell to determine the position of a generated particle by dividing the cell with a distance of $L_{\text{cp}} = 0.01 L_{\text{Cell}}$. However, not all the candidate position points are suitable for generating fluid particles. In order to avoid generating a fluid particle too close to its neighboring particles, the points located within the distance of l_0 from the existing neighboring particles are excluded from the candidate positions, as the unfeasible region shown in Fig. 5.20. In the remaining candidate position points, the one that has the farthest perpendicular distance from the outlet boundary is chosen as the point for particle generation. If no candidate point remains, the fluid particle generation procedure stops, and if there are several candidate points keeping the same largest distance, the center one is chosen as the new particle position. The velocity of the new fluid particle is calculated on the basis of mass conservation in each

Figure 5.19 Karman vortices simulated with improved outflow boundaries ($Re = 100$) [36].

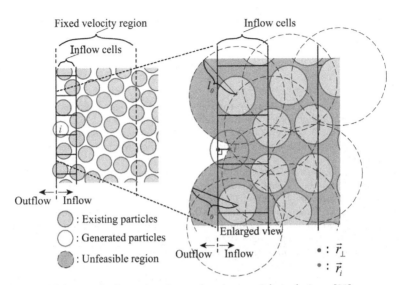

Figure 5.20 Inlet–outlet boundary in overlapping particle technique [38].

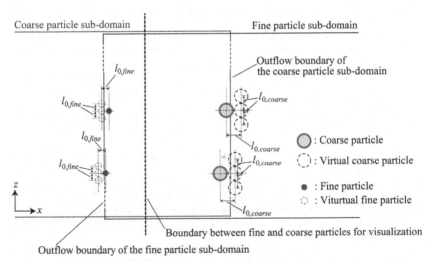

Figure 5.21 Virtual particles outside inlet–outflow boundaries [38].

subdomain, and the pressure is interpolated from the pressure of particles in other subdomains. This position selection process for generating new particles is a little time-consuming.

At the outflow boundaries of the fine particle subdomain and coarse particle subdomain, the virtual particles are imaged to supplement the PND deficiency, as shown in Fig. 5.21. The pressure of the virtual particles is set from the interpolation of pressure obtained in the other subdomains.

References

[1] S. Koshizuka, Y. Oka, Moving-particle semi-implicit method for fragmentation of incompressible fluid, Nucl. Sci. Eng. 123 (3) (1996) 421−434.

[2] S. Koshizuka, A. Nobe, Y. Oka, Numerical analysis of breaking waves using the moving particle semi-implicit method, Int. J. Numer. Methods Fluids 26 (7) (1998) 751−769.

[3] Z. Zheng, G. Duan, N. Mitsume, S. Chen, S. Yoshimura, A novel ghost cell boundary model for the explicit moving particle simulation method in two dimensions, Comput. Mech. 66 (1) (2020) 87−102.

[4] Z. Zheng, G. Duan, N. Mitsume, S. Chen, S. Yoshimura, An explicit MPS/FEM coupling algorithm for three-dimensional fluid-structure interaction analysis, Eng. Anal. Bound. Elem. 121 (2020) 192−206.

[5] S.J. Cummins, M. Rudman, An SPH projection method, J. Comput. Phys. 152 (2) (1999) 584−607.

[6] N. Mitsume, S. Yoshimura, K. Murotani, T. Yamada, Explicitly represented polygon wall boundary model for the explicit MPS method, Comput. Part. Mech. 2 (1) (2015) 73−89.

[7] H. Akimoto, Numerical simulation of the flow around a planing body by MPS method, Ocean Eng. 64 (2013) 72−79.

[8] T. Matsunaga, K. Shibata, K. Murotani, S. Koshizuka, Fluid flow simulation using MPS method with mirror particle boundary representation, Trans. Jpn. Soc. Comput. Eng. Sci. (2016) 20160002 (in Japanese).

[9] N. Trask, M. Maxey, X. Hu, Compact moving least squares: an optimization framework for generating high-order compact meshless discretizations, J. Comput. Phys. 326 (2016) 596−611.

[10] W. Hu, N. Trask, X. Hu, W. Pan, A spatially adaptive high-order meshless method for fluid−structure interactions, Computer Methods Appl. Mech. Eng. 355 (2019) 67−93.

[11] T. Tamai, S. Koshizuka, Least squares moving particle semi-implicit method, Comput. Part. Mech. 1 (3) (2014) 277−305.

[12] X. Jin, G. Li, N.R. Aluru, Positivity conditions in meshless collocation methods, Computer Methods Appl. Mech. Eng. 193 (12−14) (2004) 1171−1202.

[13] B. Seibold, Minimal positive stencils in meshfree finite difference methods for the Poisson equation, Computer Methods Appl. Mech. Eng. 198 (3−4) (2008) 592−601.

[14] T. Harada, S. Koshizuka, Improvement of wall boundary calculation model for MPS method, Trans. Jpn. Soc. Comput. Eng. Sci. (2008) 2008006 (in Japanese).

[15] T. Zhang, S. Koshizuka, K. Murotani, K. Shibata, E. Ishii, M. Ishikawa, Improvement of boundary conditions for non-planar boundaries represented by polygons with an initial particle arrangement technique, Int. J. Comput. Fluid Dyn. 30 (2) (2016) 155−175.

[16] T. Zhang, S. Koshizuka, K. Murotani, K. Shibata, E. Ishii, Improvement of pressure distribution to arbitrary geometry with boundary condition represented by polygons in particle method, Int. J. Numer. Methods Eng. 112 (7) (2017) 685−710.

[17] T. Zhang, S. Koshizuka, P. Xuan, J. Li, C. Gong, Enhancement of stabilization of MPS to arbitrary geometries with a generic wall boundary condition, Computers Fluids 178 (2019) 88−112.

[18] T. Matsunaga, K. Shibata, S. Koshizuka, Boundary integral based polygon wall representation in the MPS method, Trans. JSME (Japanese) 84 (864) (2018) 18−00197.
[19] T. Matsunaga, N. Yuhashi, K. Shibata, S. Koshizuka, A wall boundary treatment using analytical volume integrations in a particle method, Int. J. Numer. Methods Eng. 121 (18) (2020) 4101−4133.
[20] T. Matsunaga, A. Södersten, K. Shibata, S. Koshizuka, Improved treatment of wall boundary conditions for a particle method with consistent spatial discretization, Computer Methods Appl. Mech. Eng. 358 (2020) 112624.
[21] G. Duan, T. Matsunaga, A. Yamaji, S. Koshizuka, M. Sakai, Imposing accurate wall boundary conditions in corrective-matrix-based moving particle semi-implicit method for free surface flow, Int. J. Numer. Methods Fluids 93 (1) (2021) 148−175.
[22] M. Tanaka, R. Cardoso, H. Bahai, Multi-resolution MPS method, J. Comput. Phys. 359 (2018) 106−136.
[23] M. Tanaka, T. Masunaga, Stabilization and smoothing of pressure in MPS method by quasi-compressibility, J. Comput. Phys. 229 (11) (2010) 4279−4290.
[24] G.A. Dilts, Moving least squares particle hydrodynamics II: conservation and boundaries, Int. J. Numer. Methods Eng. 48 (10) (2000) 1503−1524.
[25] S. Marrone, A. Colagrossi, D. Le Touzé, G. Graziani, Fast free-surface detection and level-set function definition in SPH solvers, J. Comput. Phys. 229 (10) (2010) 3652−3663.
[26] C. Sun, Z. Shen, M. Zhang, Surface treatment technique of MPS method for free surface flows, Eng. Anal. Bound. Elem. 102 (2019) 60−72.
[27] K. Shibata, I. Masaie, M. Kondo, K. Murotani, S. Koshizuka, Improved pressure calculation for the moving particle semi-implicit method, Comput. Part. Mech. 2 (1) (2015) 91−108.
[28] G. Duan, S. Koshizuka, A. Yamaji, B. Chen, X. Li, T. Tamai, An accurate and stable multiphase moving particle semi-implicit method based on a corrective matrix for all particle interaction models, Int. J. Numer. Methods Eng. 115 (10) (2018) 1287−1314.
[29] G. Duan, T. Matsunaga, S. Koshizuka, A. Yamaguchi, M. Sakai, New insights into error accumulation due to biased particle distribution in semi-implicit particle methods, Computer Methods Appl. Mech. Eng. 388 (2022) 114219.
[30] X. Chen, G. Xi, Z.-G. Sun, Improving stability of MPS method by a computational scheme based on conceptual particles, Computer Methods Appl. Mech. Eng. 278 (2014) 254−271.
[31] N. Tsuruta, A. Khayyer, H. Gotoh, Space potential particles to enhance the stability of projection-based particle methods, Int. J. Comput. Fluid Dyn. 29 (1) (2015) 100−119.
[32] E. Toyota, A particle method with variable spatial resolution for incompressible flows, in: Proceedings of the 19th Symposium on Computational Fluid Dynamics, 2005, 2005.
[33] T. Matsunaga, S. Koshizuka, T. Hosaka, E. Ishii, Moving surface mesh-incorporated particle method for numerical simulaiton of a liquid droplet, J. Comput. Phys. 409 (2020) 109349.
[34] Z. Wang, T. Sugiyama, On the free surface boundary of moving particle semi-implicit method for thermocapillary flow, Eng. Anal. Bound. Elem. 135 (2022) 266−283.
[35] A. Shakibaeinia, Y.C. Jin, A weakly compressible MPS method for modeling of open-boundary free-surface flow, Internat. J. Numer. Methods Fluids 63 (10) (2010) 1208−1232.

[36] K. Shibata, S. Koshizuka, K. Murotani, M. Sakai, I. Masaie, Boundary conditions for simulating Karman vortices using the MPS method, Jpn. Soc. Simul. Technol. 2 (2) (2015) 235–254.

[37] G. Duan, B. Chen, X. Zhang, Y. Wang, A multiphase MPS solver for modeling multi-fluid interaction with free surface and its application in oil spill, Comput. Methods Appl. Mech. Eng. 320 (2017) 133–161.

[38] K. Shibata, S. Koshizuka, T. Matsunaga, I. Masaie, The overlapping particle technique for multi-resolution simulation of particle methods, Comput. Methods Appl. Mech. Eng. 325 (2017) 434–462.

CHAPTER SIX

Surface tension models

Surface tension is essential in droplet formation and multiphase flow, especially at the microscale. Within the particle framework, three main surface tension models, that is, potential-based model, continuum model, and stress-based model, have been developed. In this chapter, these three types of surface tension models are introduced with examples. The surface tension is calculated as a pairwise potential in the potential-based model. As for the continuum model, surface tension is converted to a volumetric force. For these two models the surface tension is directly considered in the governing equations. As opposed to these, surface tension is incorporated in stress balance equations on the surface and serves as a boundary condition in the stress-based model.

6.1 Potential-based model

Shirakawa et al. [1] proposed a surface tension model based on particle internal potential force for moving particle semi-implicit (MPS) method, which is similar to molecular force. Every particle interacts with its neighboring particles, and the potential force is repulsive and attractive, respectively, when the particle distance is less and larger than a certain value. The force acting on the internal particles with the same physical properties is balanced for a uniform particle distribution, while for interface particles the nonbalanced potential force between different particles acts as the surface tension. Theoretically, various potential function formulations, satisfying repulsive force within a short distance and attractive force in a large distance, are applicable. Shirakawa et al. [1] proposed a curvilinear function which is repulsive force in the particle distance $[0, l_0]$ and attractive force in particle distance $[l_0, 2l_0]$, as shown in Fig. 6.1.

The force applied to the liquid particles in the vicinity of the surface is formulated as

$$f_i^{st} = \sum_{j \neq i} -\frac{\partial P(r_{ij})}{\partial r_{ij}} \frac{\mathbf{r}_{ij}}{r_{ij}} \left(\frac{n^0}{n^i}\right)^2 \tag{6.1}$$

Moving Particle Semi-implicit Method.
DOI: https://doi.org/10.1016/B978-0-443-13508-8.00006-8

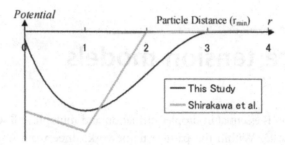

Figure 6.1 Potential for the surface tension [2].

where $P(r_{ij})$ is the potential between i and j particles, $\mathbf{r}_{ij} = \mathbf{r}_j - \mathbf{r}_i$, and $r_{ij} = |\mathbf{r}_j - \mathbf{r}_i|$.

Kondo et al. [2] also proposed a new potential function (Eqs. 6.2 and 6.3), which is much smoother than Shirakawa's function.

$$P(r_{ij}) = Cp(r_{ij}) \tag{6.2}$$

$$p(r_{ij}) = \begin{cases} \dfrac{1}{3}\left(r_{ij} - \dfrac{3}{2}l_0 + \dfrac{1}{2}r_e\right)(r_{ij} - r_e)^2 & (r_{ij} < r_e) \\ 0 & (r_{ij} \geq r_e) \end{cases} \tag{6.3}$$

where r_e is the effective radius of the potential function and C is a fitting coefficient. An empirical value of $r_e = 3.2l_0$ is used. The shape of the potential function Eq. (6.3) is shown in Fig. 6.1. The total potential E_i^{st} of particle i with respect to its neighboring particles is calculated by

$$E_i^{st} = \sum_{j \neq i} P(r_{ij}) \tag{6.4}$$

The force acting on particle i can be calculated by the differential of the potential E_i^{st} with respect to particle position vector.

$$f_i^{st} = -\frac{\partial E_i^{st}}{\partial \mathbf{r}_i} = \sum_{j \neq i} \frac{\partial P(r_{ij})}{\partial r_{ij}} \frac{\mathbf{r}_{ij}}{r_{ij}} = -\sum_{j \neq i} C(r_{ij} - l_0)(r_{ij} - r_e)\frac{\mathbf{r}_{ij}}{r_{ij}} \tag{6.5}$$

Here, the arbitrary parameter C needs to be determined before the calculation. Kondo et al. [2] built a relation with surface tension coefficient by creating a new fluid surface, as shown in Fig. 6.2. The work required to detach fluid A from fluid B with a surface of l_0^2 is estimated as

$$2\sigma l_0^2 = \sum_{i \in A, j \in B} P(r_{ij}) \tag{6.6}$$

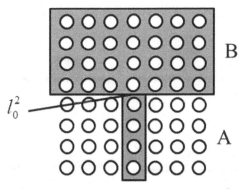

Figure 6.2 Work needed to create two new surfaces with area l_0^2[2].

where σ is the surface tension coefficient. Consequently, C can be derived as

$$C = \frac{2\sigma l_0^2}{\sum_{i \in A, j \in B, r_{ij} < r_e} \frac{1}{3} \left(r_{ij} - \frac{3}{2} l_0 + \frac{1}{2} r_e\right)(r - r_e)^2} \tag{6.7}$$

When the potential-based surface tension model is used for a fluid contacting with the solid wall, the contact angle should be considered. According to Young's relation the force balance equation at fluid−gas−solid connection is written as

$$\sigma_s - \sigma_{fs} - \sigma_f \cos\theta = 0 \tag{6.8}$$

where θ is the contact angle; σ_s, σ_{fs}, and σ_f are the surface energies of the solid surface, fluid−solid surface, and fluid surface, respectively.

To create the relation of contact angle and potential force, the following two equations are derived by detaching one fluid from the fluid−fluid interface and the fluid−solid interface, respectively.

$$2\sigma_f r_{\min}{}^2 = C_f \mathrm{P}\left(r_{\min}{}^2\right) \tag{6.9}$$

$$\left(\sigma_f + \sigma_s - \sigma_{fs}\right) r_{\min}{}^2 = C_{fs} \mathrm{P}\left(r_{\min}{}^2\right) \tag{6.10}$$

where C_f and C_{fs} are defined as the potential force coefficients between fluid particles and between the fluid particle and solid particle, respectively.

With the above equations the relation between potential force and contact angle is derived as

$$C_{fs} = \frac{C_f}{2}(1 + \cos\theta) \tag{6.11}$$

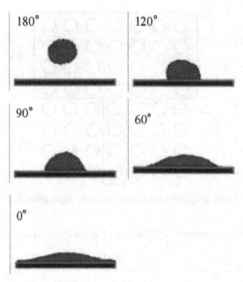

Figure 6.3 Contact angle calculation with the potential-based surface tension model [2].

The contact angle calculation using Kondo's model is shown in Fig. 6.3.

The potential-based surface tension model has the merit of easily being implemented. It does not need gas particles when calculating the tension at the free surface, which can reduce the computation cost in the practical application. However, the accuracy of the potential-based surface tension model is insufficient because it calculates the relative difference between forces in different fluids. In fact the parameter σ used in the model cannot represent the surface tension coefficient at two-fluid interface. One needs to set separate values of σ for two fluids, and a calibration process of σ values is necessary to make their difference equal to the physical surface tension. Moreover, the potential-based surface tension model calculates not only the forces of surface particles but also the forces of internal particles. In the condition of nonuniform distribution of internal particles, the forces acting on the target particle cannot be balanced. As a result, the numerical instability is usually a problem. Nevertheless, due to its simplicity and high efficiency, the potential-based surface tension model has been applied to many simulations in engineering.

Shirakawa et al. [1] simulated jet breakup in liquid pool with the potential-based surface tension model. Liu et al. [3] replaced l_0 in Eq. (6.3) by $1.5l_0$ in the simulation of liquid dispersion by hydrophobic/hydrophilic mesh packing. Hattori et al. [4] simulated dynamic contact angle of a droplet sliding on an inclined surface using the potential-based surface tension model. Li et al. [5] simulated stratification behavior of two immiscible fluids. Chen et al. [6] used the model to the simulation of melt behavior in the fuel support piece during a nuclear reactor severe accident. There are also many other studies using the potential-based surface tension model.

6.2 Continuum model

The continuum surface force (CSF) model is first proposed by Brackbill et al. [7]. Using the CSF, the surface tension is converted to a volumetric force in a transition region near the interface. Surface tension in CSF model is expressed by

$$f = \sigma \kappa \delta \mathbf{n} \tag{6.12}$$

where f is the surface tension force, κ is the curvature, σ is the surface tension coefficient, δ is the delta function, and $\mathbf{n}$ is the normal vector of the surface. By introducing the color function C, Eq. (6.12) can be converted to

$$f = \sigma \kappa \nabla C \tag{6.13}$$

C is the color function defined as

$$C_i = \begin{cases} 0 & i \text{ is in the specified phase} \\ 1 & i \text{ is in the other phase} \end{cases}$$

To avoid the discontinuity of the color function across the interface, a smoothed color function is usually adopted:

$$\tilde{C}_i = \frac{\sum_j w_j C_j}{\sum_j w_j} \tag{6.15}$$

Here, w_j is the weighted kernel for particle j in the neighbor of particle i. Using Eq. (6.15) the color function is smoothed across the interface and varies from zero to one smoothly from one phase to the other phase.

The gradient of the smoothed color function can be calculated based on the gradient model in the MPS. High-order models such as the least-square-based MPS model can be employed to improve the accuracy.

Curvature calculation is the most difficult part in the CSF model. Curvature can be calculated from the unit normal vector of the surface:

$$\kappa = - \nabla \cdot \mathbf{n} \tag{6.16}$$

The unit normal vector can be calculated by

$$\mathbf{n} = \frac{\nabla C}{|\nabla C|} \tag{6.17}$$

Although this is the most widely adopted formulation to calculate curvature, some researchers have reported that huge errors and numerical instability can be raised using this formulation.

To improve the accuracy and stability, several researchers have developed new strategies to calculate the curvature.

Liu et al. [8] developed a machine-learning-based curvature calculation for the particle-based method. 3×3 Cartesian cells are configured centered on the interface particle as shown in Fig. 6.4. The color function for each cell is calculated by weighted sum of the color function of particles around each cell center:

$$D_m = \frac{\sum_j w_{mj} \tilde{C}_j}{\sum_j w_{mj}}, m = 1, 2 \ldots 9. \tag{6.18}$$

Here, D_m is the color function of cell m. Particles located within each cell are used in the calculation of the color function of the cell.

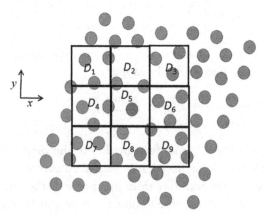

Figure 6.4 Configuration of the background cells in the machine-learning-based surface tension model [8].

Using the nondimensional curvature $\kappa_i \cdot r_e$ the goal is to find the function ψ:

$$\kappa_i \cdot r_e = \psi([D_1, D_2 \ldots D_9]). \qquad (6.19)$$

A dataset containing curvature and color functions is generated using circles of different sizes. A 1×1 domain is resolved by 1000×1000 particles, and the interface with the shape of varied circles is configured. The radius of circles ranges from 0.005 to 0.5 with the increment equal to 0.001. The dataset of color functions and curvature are generated near the interface along each circle. In total, 472,340 sets of color function and curvature data are generated. Among these datasets, 70% is used for training, 15% is used for validation, and 15% is used for test.

KERAS, an open source library, is used to build the neural network. A sequential model is employed. The input is a 9-dimensional array of the color function. Several tests using different cell sizes, hidden layers, and epochs are conducted, and the results of correlation coefficient between the test data and the data predicted by machine learning are summarized in Tables 6.1 and 6.2. Using a cell size equal to $4l_0$, two hidden layers with 500 epochs would give enough accuracy. A test is conducted as well to demonstrate the advantage of using smoothed color function in calculating the color function of the cells, and the result is summarized in Table 6.3. The accuracy is improved significantly using smoothed color function. Fig. 6.5 shows the structure of the network. Two hidden layers with 100 neurons are used. The output is the

Table 6.1 Result of the tests using different cell sizes [8].

Cell size	Layer	Epoch	Correlation coefficient
$3l_0$	2	500	0.8271
$4l_0$	2	500	0.9722
$5l_0$	2	500	0.9554
$6l_0$	2	500	0.9623

Table 6.2 Result of the tests using different layers and epochs.

Layer	Epoch	Correlation coefficient
1	500	0.9574
2	500	0.9722
3	500	0.9738
2	250	0.9696
2	1000	0.9734

Table 6.3 Result of the tests with smooth and without smooth.

Smooth	Layer	Epoch	Correlation coefficient
Yes	2	500	0.9722
No	2	500	0.8552

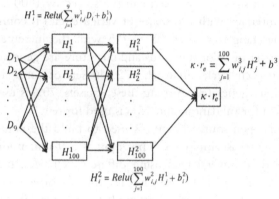

Figure 6.5 Structure of the network used in training the machine learning model.

normalized nondimensional curvature. *ReLU* activation function is used for the hidden layer, and linear activation function is used for the output.

To test the curvature estimation using machine learning the curvature along a circle and the curve defined by a sine wave, $y(x) = A\sin(x)$, is calculated and compared with the analytical result. Calculations using the derivative model are also conducted for comparison. In the derivative model, the divergence model with a corrective matrix is used to calculate the curvature from unit interface normal:

$$\kappa = -\frac{d}{n^0} \sum_{j \neq i} \frac{\mathbf{n}_j - \mathbf{n}_i}{|\mathbf{r}_{ij}|^2} \cdot M_i \mathbf{r}_{ij} w_{ij} \tag{6.20}$$

where M_i is the corrective matrix:

$$M_i = \begin{bmatrix} \frac{d}{n^0} \sum \frac{w_{ij}x_{ij}^2}{|\mathbf{r}_{ij}|^2} & \frac{d}{n^0} \sum \frac{w_{ij}x_{ij}y_{ij}}{|\mathbf{r}_{ij}|^2} \\ \frac{d}{n^0} \sum \frac{w_{ij}x_{ij}y_{ij}}{|\mathbf{r}_{ij}|^2} & \frac{d}{n^0} \sum \frac{w_{ij}y_{ij}^2}{|\mathbf{r}_{ij}|^2} \end{bmatrix}^{-1}, \tag{6.21}$$

In the calculation of the curvature of a sine wave, $A = 1.0$ is used. The calculation domain is $2\pi \times 2\pi$ domain divided by the sine wave.

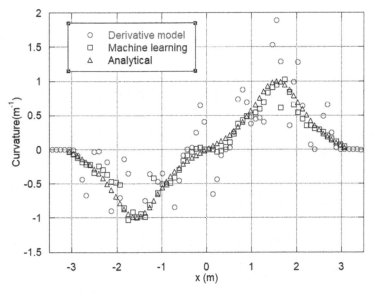

Figure 6.6 Comparison of the curvature of the sine wave predicted by machine learning and analytical results [8].

100×100 particles are used. Comparison of the predicted result using the derivative model, machine learning, and analytical results is shown in Fig. 6.6. The derivative model shows large fluctuation, while using machine learning, comparable results are obtained.

To accurately calculate curature, Duan et al. [9] also proposed a contoured continuum surface force (CCSF) model. In their model the smoothed color function f at an arbitrary position (x, y) near the reference particle i can be calculated from the color function C through a smoothing technique.

$$f(x, y) = \frac{\sum_j G(x - x_j, y - y_j, r_s) C_j}{\sum_j G(x - x_j, y - y_j, r_s)} \tag{6.22}$$

where r_s is the smoothed radius and G is the Gaussian kernel. The local contour passing through particle i is actually $f(x, y) = f(x_i, y_i)$, from which directly calculating the curvature is relatively difficult. Thus f is expanded by the Taylor series. To maintain high accuracy of expansion the partial derivatives of f are calculated analytically. f can be rewritten as follows:

$$f(x, y) = \frac{h(x, y)}{g(x, y)} \tag{6.23}$$

where

$$h(x, y) = \sum_j G(x - x_j, y - y_j, r_s) C_j \tag{6.24}$$

$$g(x, y) = \sum_j G(x - x_j, y - y_j, r_s) \tag{6.25}$$

It is convenient to compute the first- and second-order partial derivatives of h and g at point (x_i, y_i), as shown in the following:

$$h_{x,i} = \sum_j C_j \frac{\partial G(x - x_j, y - y_j, r_s)}{\partial x} \quad g_{x,i} = \sum_j \frac{\partial G(x - x_j, y - y_j, r_s)}{\partial x}$$

$$h_{y,i} = \sum_j C_j \frac{\partial G(x - x_j, y - y_j, r_s)}{\partial y} \quad g_{y,i} = \sum_j \frac{\partial G(x - x_j, y - y_j, r_s)}{\partial y}$$

$$h_{xx,i} = \sum_j C_j \frac{\partial^2 G(x - x_j, y - y_j, r_s)}{\partial x^2} \quad g_{xx,i} = \sum_j \frac{\partial^2 G(x - x_j, y - y_j, r_s)}{\partial x^2}$$

$$\tag{6.26}$$

$$h_{xy,i} = \sum_j C_j \frac{\partial^2 G(x - x_j, y - y_j, r_s)}{\partial x \partial y} \quad g_{xy,i} = \sum_j \frac{\partial^2 G(x - x_j, y - y_j, r_s)}{\partial x \partial y}$$

and

$$h_{yy,i} = \sum_j C_j \frac{\partial^2 G(x - x_j, y - y_j, r_s)}{\partial y^2} \quad g_{yy,i} = \sum_j \frac{\partial^2 G(x - x_j, y - y_j, r_s)}{\partial y^2} \tag{6.27}$$

Then the first- and the second-order partial derivatives of f at point (x_i, y_i) can be calculated from the derivatives of g and h according to the derivative rules. These partial derivatives are

$$f_{x,i} = \frac{h_{x,i}g_i - g_{x,i}h_i}{g_i^2}$$

$$f_{y,i} = \frac{h_{y,i}g_i - g_{y,i}h_i}{g_i^2}$$

$$f_{xx,i} = \frac{(h_{xx,i}g_i - g_{xx,i}h_i)g_i - 2(h_{x,i}g_{x,i}g_i - g_{x,i}g_{x,i}h_i)}{g_i^3} \tag{6.28}$$

$$f_{yy,i} = \frac{(h_{yy,i}g_i - g_{yy,i}h_i)g_i - 2(h_{y,i}g_{y,i}g_i - g_{y,i}g_{y,i}h_i)}{g_i^3}$$

$$f_{xy,i} = \frac{(h_{xy,i}g_i - g_{xy,i}h_i - h_{x,i}g_{y,i} - h_{y,i}g_{x,i})g_i + 2g_{x,i}g_{y,i}h_i)}{g_i^3}$$

After obtaining the partial derivatives of f, f can be expanded at (x_i, y_i) according to the Taylor series expansion:

$$f(x,y) = f(x_i,y_i) + f_{x,i}(x - x_i) + f_{y,i}(y - y_i) + \frac{1}{2}f_{xx,i}(x - x_i)^2 + \frac{1}{2}f_{yy,i}(y - y_i)^2$$
$$+ f_{xy,i}(x - x_i)(y - y_i) + o(r_s^3)$$

(6.29)

The local contour function of f passing through (x_i, y_i) is obtained by setting the following:

$$f(x,y) = f(x_i, y_i) \qquad (6.30)$$

When the high-order error term $[o(r_s^3)]$ is omitted in Eq. (6.29) the local contour expression can be expressed as

$$f_{x,i}(x - x_i) + f_{y,i}(y - y_i) + \frac{1}{2}f_{xx,i}(x - x_i)^2 + \frac{1}{2}f_{yy,i}(y - y_i)^2$$
$$+ f_{xy,i}(x - x_i)(y - y_i) = 0$$

(6.31)

If y is regarded as a function of x in Eq. (6.31), its curvature can be analytically calculated according to the following definition:

$$\kappa_i = \frac{y_i''}{(1 + y_i'^2)^{3/2}} = \frac{2(f_{x,i}f_{y,i}f_{xy,i} - \frac{1}{2}f_{y,i}f_{y,i}f_{xx,i} - \frac{1}{2}f_{x,i}f_{x,i}f_{yy,i})}{(f_{x,i}^2 + f_{y,i}^2)^{3/2}} \qquad (6.32)$$

Another advantage of CCSF is that it can be easily extended into three-dimensional (3D) simulations. Specifically, the derivatives of a smoothed color function with respect to variable z can be calculated similarly. The mean curvature of the local contour surface in 3D can also be derived as follows:

$$\kappa_i = \frac{f_{x,i}f_{y,i}f_{xy,i} + f_{x,i}f_{z,i}f_{xz,i} + f_{y,i}f_{z,i}f_{yz,i}}{(f_{x,i}^2 + f_{y,i}^2 + f_{z,i}^2)^{3/2}}$$
$$- \frac{f_{x,i}f_{x,i}(f_{yy,i} + f_{zz,i}) + f_{y,i}f_{y,i}(f_{xx,i} + f_{zz,i}) + f_{z,i}f_{z,i}(f_{xx,i} + f_{yy,i})}{2(f_{x,i}^2 + f_{y,i}^2 + f_{z,i}^2)^{3/2}} \qquad (6.33)$$

6.3 Stress-based model

A rather accurate approach has been developed by Matsunaga et al. [10] for modeling the surface tension at a free surface boundary. In this model, the surface tension is considered as the free-surface boundary condition rather than being treated as volume forces.

On a fluid interface the following stress balance equations are obtained [7]:

$$(P_1 - P_2 + \sigma\kappa)\mathbf{n}_i = (\varepsilon_{1ik} - \varepsilon_{2ik}) \cdot \mathbf{n}_k + \frac{\partial \sigma}{\partial x_i} \tag{6.34}$$

where $\mathbf{n}_i$ is the unit normal vector (into fluid 2) at the interface, P_α is the pressure, α denotes each of fluids and is defined as $\alpha = 1, 2$, and $\varepsilon_{\alpha ik}$ denotes the viscous tension tensor, which is defined as

$$\varepsilon_{\alpha ik} = \mu_\alpha \left(\frac{\partial u_i}{\partial x_k} + \frac{\partial u_k}{\partial x_i} \right) \tag{6.35}$$

In the case of the free-surface boundary, gas density is assumed to be infinitesimal; thus $P_{gas} = 0$ and $\varepsilon_{gas\_ik} = 0$. If Eq. (6.34) is projected normal to the surface, the boundary condition for pressure is obtained as

$$(P + \sigma\kappa)\mathbf{n} = \mu\left(\nabla \otimes \mathbf{u} + (\nabla \otimes \mathbf{u})^T\right)\mathbf{n} \tag{6.36}$$

If Eq. (6.34) is projected parallel to the surface, the boundary condition for velocity is written as

$$\mu(\mathbf{I}\text{-}\mathbf{n} \otimes \mathbf{n}) \cdot \left(\nabla \otimes \mathbf{u} + (\nabla \otimes \mathbf{u})^T\right) \cdot \mathbf{n} = (\mathbf{I}\text{-}\mathbf{n} \otimes \mathbf{n}) \cdot \nabla\sigma \tag{6.37}$$

If the surface tension is assumed constant, the right side of Eq. (6.37) equals zero. Nevertheless, this equation is only valid for a viscous fluid.

As mentioned in Section 5.2.3, the free-surface boundary can be explicitly represented by a moving surface mesh. As illustrated in Fig. 6.7, the surface mesh consists of discrete nodes, called free surface nodes. These surface nodes are strongly involved in the simulations by providing the boundary conditions for the nearby fluid particles. Since the free-surface boundary is explicitly given, the free-surface fluctuations caused by the conventional free-surface detection method are prevented. Surface nodes will be inserted or deleted using the polynomial interpolation to deal with the free-surface deformation, as shown in Fig. 6.8.

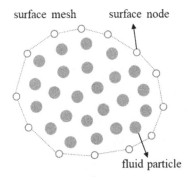

Figure 6.7 Schematic illustration of the moving surface mesh, consisting of surface nodes [10].

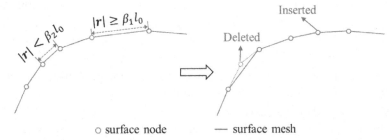

Figure 6.8 Surface mesh refinement [10].

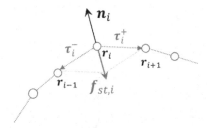

Figure 6.9 Surface tension force vector ($\mathbf{f}_{st,i}$) at surface node i [11].

As illustrated in Fig. 6.9, the surface tension force $\mathbf{f}_{st,i}$ on a surface node i is directly calculated based on the geometry of the surface mesh [11]:

$$\mathbf{f}_{st,i} = \sigma \kappa \mathbf{n}_i = \frac{\sigma}{S_i} \left(\boldsymbol{\tau}_i^- + \boldsymbol{\tau}_i^+ \right) \qquad (6.38)$$

$$S_i = \frac{|\mathbf{r}_{i-1} - \mathbf{r}_i| + |\mathbf{r}_{i+1} - \mathbf{r}_i|}{2} \qquad (6.39)$$

$$\tau_i^- = \frac{\mathbf{r}_{i-1} - \mathbf{r}_i}{|\mathbf{r}_{i-1} - \mathbf{r}_i|} \qquad\qquad (6.40)$$

$$\tau_i^+ = \frac{\mathbf{r}_{i+1} - \mathbf{r}_i}{|\mathbf{r}_{i+1} - \mathbf{r}_i|} \qquad\qquad (6.41)$$

where $\mathbf{n}_i$ indicates an outward unit normal vector to the free surface, S_i denotes the surface area occupied by node i, and τ_i^- and τ_i^+ are the unit tangent vectors pointing to the adjacent nodes $i - 1$ and $i + 1$, respectively.

The oscillatory motion of a square droplet has been accurately predicted with this method (see Fig. 6.10). A homogeneous pressure field was finally achieved, agreeing with the analytical value (Laplace pressure). In the case of temperature dependence, the surface tension variation can be considered in Eq. (6.37). The resulting phenomenon is called Marangoni-induced convection or thermocapillary convection. As shown in Fig. 6.11, the Marangoni convection in an open cavity has been simulated by Wang and Sugiyama [12]. Compared with previously reported results [13], good agreements were obtained.

Duan and Sakai [14] extended this methodology by carefully detecting the free-surface particles [15] rather than adopting the moving surface mesh. As such, node connectivity is no longer needed (see Fig. 6.12). The surface normal is calculated based on a color function, which is

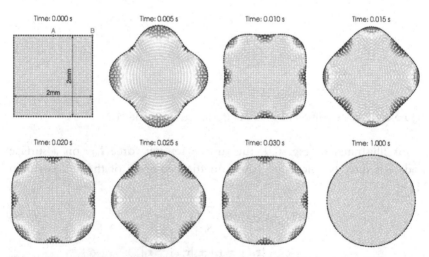

Figure 6.10 Snapshots of square droplet oscillation, colored by pressure [10].

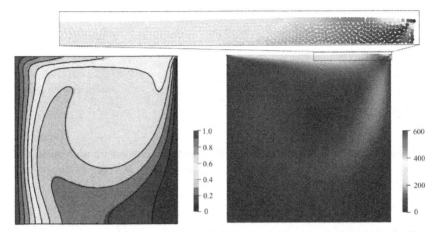

Figure 6.11 Dimensionless temperature (left) and velocity magnitude (right) distributions of Marangoni convection in an open cavity [12].

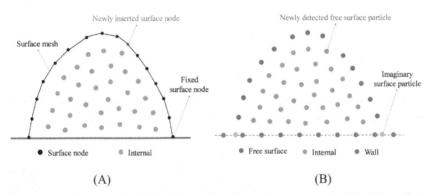

Figure 6.12 Comparison of the free-surface boundary represented by (A) surface mesh and (B) detected free-surface particles [14].

defined as

$$
\chi_i = \begin{cases}
1, & i \in \text{ free surface particles} \\
0, & i \in \text{ vicinity or internal particles} \\
\frac{\partial \chi}{\partial n}\big|_i = 0, & i \in \text{ wall particles}
\end{cases}
\tag{6.42}
$$

The condition $\frac{\partial \chi}{\partial n}\big|_i = 0$ indicates a free-surface normal parallel to the wall. The surface normal equals the normalization of $\nabla \chi$. For convenience a new coordinate system is built on particle i by selecting the

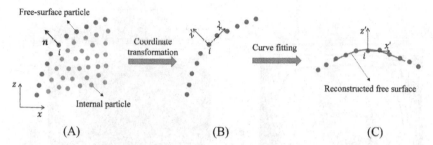

Figure 6.13 Reconstruction of the free surface via the coordinate transformation and curving fitting methods [14]. (A) Original coordinate system. (B) Free-surface particles. (C) New coordinate system.

z-axis in the surface normal $\mathbf{n}_i$, as illustrated in Fig. 6.13. The curvature is estimated based on the free-surface curving fitting, which can be written as

$$z(x) = A + Bx + Cx^2 \text{ for 2D} \tag{6.43}$$

$$z(x, y) = A + Bx + Cy + Dx^2 + Ey^2 + Fxy \text{ for 3D} \tag{6.44}$$

The coefficients A to F in Eqs. (6.43) or (6.44) are obtained by the least square method [16]. Consequently the curvature is directly estimated as

$$\kappa_i = \frac{2C}{(1+B)^{3/2}} \text{ for 2D} \tag{6.45}$$

$$\kappa_i = 2\frac{\left(1+B^2\right)\cdot E + \left(1+C^2\right)\cdot D - B\cdot C\cdot F}{\left(1+B^2+C^2\right)^{3/2}} \text{ for 3D} \tag{6.46}$$

It could be relatively easy to deal with topological changes and 3D problems since the surface mesh is no longer adopted. Fig. 6.14 shows the process of a 3D droplet deformation and breakup problem from a faucet [14]. The key features of the breakup phenomena have been well captured. Furthermore, the wettability calculation can be easily simulated by considering the imaginary surface particle on the wall. Fig. 6.15 illustrates a 3D cubic droplet spreading (contact angle 90 degrees) on a wall [14].

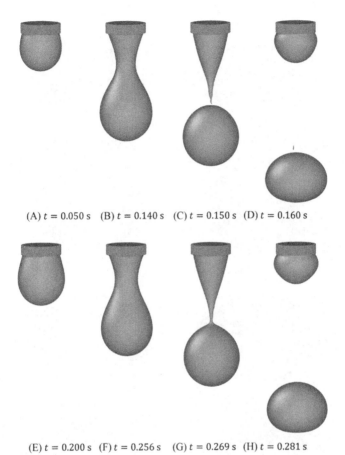

(A) $t = 0.050$ s   (B) $t = 0.140$ s   (C) $t = 0.150$ s   (D) $t = 0.160$ s

(E) $t = 0.200$ s   (F) $t = 0.256$ s   (G) $t = 0.269$ s   (H) $t = 0.281$ s

Figure 6.14 3D droplet deformation and breakup [14].

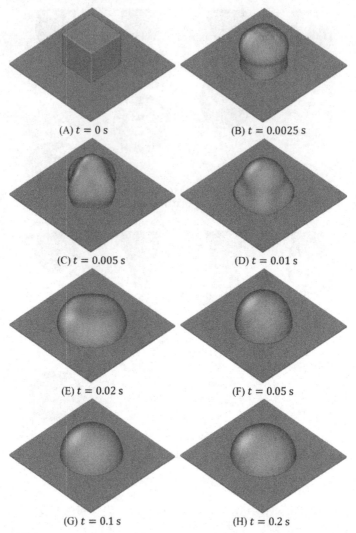

(A) $t = 0$ s   (B) $t = 0.0025$ s

(C) $t = 0.005$ s   (D) $t = 0.01$ s

(E) $t = 0.02$ s   (F) $t = 0.05$ s

(G) $t = 0.1$ s   (H) $t = 0.2$ s

Figure 6.15 Evolution of a 3D cubic droplet spreading on a wall (contact angle 90 degrees) [14].

References

[1] Noriyuki Shirakawa, Hideki Horie, Yuichi Yamamoto, Shigeaki Tsunoyama, Analysis of the void distribution in a circular tube with the two-fluid particle interaction method, J. Nucl. Sci. Technol. 38 (6) (2001) 392−402.
[2] M. Kondo, S. Koshizuka, T. Suzuki, Surface tension model using inter-particle force in particle method, in: 5th Joint ASME/JSME Fluids Engineering Conference, San Diego, California USA, 2007.

[3] Qi-xin Liu, Zhong-guo Sun, Yi-jie Sun, Xiao Chen, Guang Xi, Numerical investi-
    gation of liquid dispersion by hydrophobic/hydrophilic mesh packing using particle
    method, Chem. Eng. Sci. 202 (2019) 447−461.

[4] Tsuyoshi Hattori, Masaharu Sakai, Shigeru Akaike, Seiichi Koshizuka, Numerical
    simulation of droplet sliding on an inclined surface using moving particle semi-
    implicit method, Comput. Part. Mech. 5 (2018) 477−491.

[5] Gen Li, Yoshiaki Oka, Masahiro Furuya, Masahiro Kondo, Experiments and MPS
    analysis of stratification behavior of two immiscible fluids, Nucl. Eng. Des. 265
    (2013) 210−221.

[6] Lie Ronghua Chen, Kailun Chen, Akifumi Guo, Masahiro Yamaji, Wenxi Furuya,
    G.H. Tian, et al., Numerical analysis of the melt behavior in a fuel support piece of
    the BWR by MPS, Ann. Nucl. Energy 102 (2017) 422−439.

[7] J.U. Brackbill, D.B. Kothe, C. Zemach, A continuum method for modelling surface
    tension, J. Comp. Phys. 100 (2) (1992) 335−354.

[8] X. Liu, K. Morita, S. Zhang, Machine-learning-based surface tension model for mul-
    tiphase flow simulation using particle method, Int. J. Numer. Methods Fluids 93 (2)
    (2021) 356−368.

[9] G. Duan, S. Koshizuka, B. Chen, A contoured continuum surface force model for
    particle methods, J. Comp. Phys. 298 (2015) 280−304.

[10] T. Matsunaga, S. Koshizuka, T. Hosaka, E. Ishii, Moving surface mesh-incorporated
    particle method for numerical simulation of a liquid droplet, J. Comp. Phys. 409
    (2020) 109349.

[11] G. Tryggvason, R. Scardovelli, S. Zaleski, Direct Numerical Simulations of
    Gas−Liquid Multiphase Flows, Cambridge University Press, 2011.

[12] Z. Wang, T. Sugiyama, On the free surface boundary of moving particle semi-
    implicit method for thermocapillary flow, Eng. Anal. Bound. Elem. 135 (2022)
    266−283.

[13] B.M. Carpenter, G. Homsy, High Marangoni number convection in a square cavity:
    part II, Phys. Fluids A Fluid Dyn. 2 (2) (1990) 137−149.

[14] G. Duan, M. Sakai, An enhanced semi-implicit particle method for simulating the
    flow of droplets with free surfaces, Comp. Methods Appl. Mech. Eng. (2021)
    114338.

[15] G. Duan, T. Matsunaga, S. Koshizuka, A. Yamaguchi, M. Sakai, New insights into
    error accumulation due to biased particle distribution in semi-implicit particle meth-
    ods, Comp. Methods Appl. Mech. Eng. 388 (2022) 114219.

[16] M. Zhang, Simulation of surface tension in 2D and 3D with smoothed particle
    hydrodynamics method, J. Comp. Phys. 229 (19) (2010) 7238−7259.

CHAPTER SEVEN

Multiphase flow and turbulence models

Multiphase and turbulence flows are widely encountered in nuclear engineering, ocean engineering, and other industrial applications. The most challenging issues in numerical simulations of multiphase and turbulence flows are the capturing of largely deformed free surfaces and interfaces and the physics discontinuity at phase interfaces, including density and viscosity. Particle methods have inherent advantages in interface capturing by using different particle types. They do not need additional procedures to reconstruct the interface and surface. However, the discontinuities of density and viscosity at the phase interface need special treatments. In this chapter, the numerical models for gas—liquid, solid—liquid, and turbulence flows are introduced. In the part of gas—liquid two-phase flow, special attention is paid on the methods dealing with physics discontinuity, including two-step pressure calculation algorithm, and smoothing techniques of physical properties. In the part of solid—liquid two-phase flow, the models of discrete description and continuum description of the solid granular flows are introduced. Finally, the turbulence model in terms of large eddy simulation is introduced.

7.1 Gas—liquid two-phase flow

An inevitable problem in gas—liquid two-phase simulation is the discontinuities of fluid density and viscosity at interfaces. The discontinuities in physics may cause mathematical discontinuity and accordingly a discontinuous acceleration field at phase interfaces. These discontinuities may lead to instabilities and even end-up of calculations. Researchers have made some efforts to solve the discontinuity problems that arise from density and viscosity differences between two different particles. These approaches can be classified into the categories of two-step pressure calculation algorithm and interface smoothing schemes.

Moving Particle Semi-implicit Method.
DOI: https://doi.org/10.1016/B978-0-443-13508-8.00007-X

7.1.1 Two-step pressure calculation algorithm

The sharp density jump across the phase interface will eventually result in a divergence of pressure calculation. To avoid these problems, Koshizuka et al. [1] and Ikeda et al. [2] developed a new algorithm to calculate the gas and liquid pressure. They separated the pressure calculation routine into two steps with respect to the heavy particles (melt and water) and the light particles (vapor) to maintain the numerical stability. In the first step, the pressure of liquid particles is calculated by regarding the interface with vapor as the free surface, and then the pressure of gas particles is solved in the second step by regarding liquid particles as rigid wall. Similarly, without using density smooth technique, Shimizu et al. [3] proposed other approaches to deal with the large density ratio at air−water two-phase flow based on the concept of space potential particles (see Fig. 7.1). In fact, the concept of space potential particles is used to solve the unphysical particle overlapping that results from the continuity equation not being strictly solved to free surface particles and void space interactions.

As illustrated in Fig. 7.2, the pressure calculation is divided into two steps. The pressure of air particles is calculated by considering the velocity of water particles as the velocity boundary; then, the pressure of air particles is mirrored to the space potential particles of water particles in the vicinity of the interface, and serves as the pressure boundary in the calculation of water flow. The proposed method is coupled with some stability and accuracy improvement techniques in the verifications, for example, sloshing flow and oscillating droplet under a central force field. Khayyer et al. [4] applied the optimized particle shifting (OPS) technique to multiphase flow with a large density ratio. The particle shifting process consists of two steps as well. First, particle shifting is carried out for only the heavy phase, where OPS scheme is applied for the interface. Then, the light phase is shifted at the interface by using the PS technique.

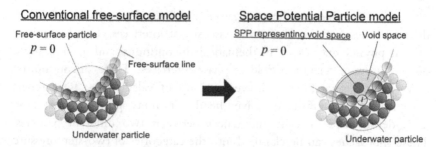

Figure 7.1 Sketch for the phase potential particle model [3].

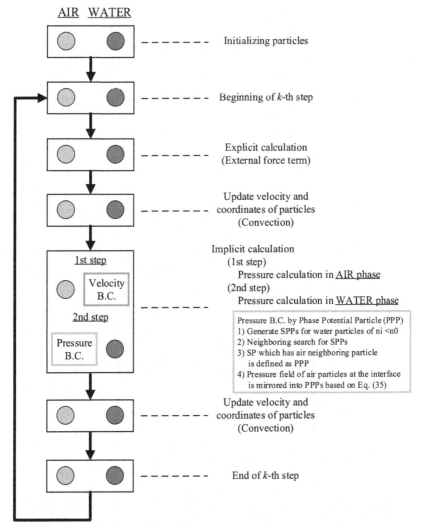

Figure 7.2 Algorithm of multiphase flow simulation [3].

The methods mentioned above do not use any density smoothing technique and turning parameters at the phase interface. With these methods, the material discontinuity and pressure continuity at the interface are kept.

7.1.2 Smoothing techniques of physical properties

The most common approach dealing with the discontinuities at the phase interface is the smoothing techniques of density and viscosity. The smoothing techniques are implemented by using spatial average formula. Shakibaeinia and Jin [5] developed a method for multiphase flow simulation by treating the system as a multidensity multiviscosity fluid, in which the weakly compressible formulations are used to solve a single set of equations for all phases. The density used in pressure calculation equation is smoothed, and different viscosity average schemes are evaluated. The pairwise particle collision model is also modified on the basis of Koshizuka's collision model by considering the multiphase force that arises due to density and viscosity differences.

In the calculation, the density at particle i position is smoothed by

$$\langle \rho \rangle_i = \frac{\sum_{j \neq i} \rho_j w\left(r_{ij}, r_e\right)}{\sum_{j \neq i} w\left(r_{ij}, r_e\right)} = \frac{1}{\langle n \rangle_i} \sum_{j \neq i} \rho_j w\left(r_{ij}, r_e\right) \tag{7.1}$$

As shown in Fig. 7.3, the density gradually changes between phase I and phase II at the transition region, whereas it is equal to the real fluid density at a position far from the interface.

With the density smoothing, the pairwise particle collision model for the multiphase system is written as

$$\begin{cases} \mathbf{u}'_i = \mathbf{u}_i - \dfrac{1}{\rho_i}\left((1+\varepsilon)\dfrac{\rho_i \rho_j}{\rho_i + \rho_j}\mathbf{u}^n_{ij} \right) \\[4mm] \mathbf{u}'_j = \mathbf{u}_j + \dfrac{1}{\rho_j}\left((1+\varepsilon)\dfrac{\rho_i \rho_j}{\rho_i + \rho_j}\mathbf{u}^n_{ij} \right) \end{cases} \tag{7.2}$$

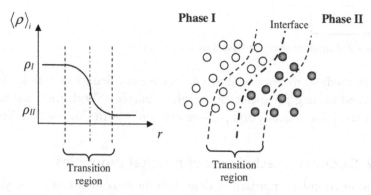

Figure 7.3 Sketch of smoothed density at the phase interface [5].

where ε is the collision ratio. The relative velocity of the particles is given by

$$\mathbf{u}_{ij}^{n} = \left(\mathbf{u}_{ij} \cdot \mathbf{e}_{ij}\right)\mathbf{e}_{ij} \qquad (7.3)$$

Likewise, the viscosity discontinuity across the interface yields a multiphase force due to the nonzero viscous gradient. The viscous term in the moving particle semi-implicit (MPS) method is given by

$$\nabla_T = \nabla(\mu(\nabla \cdot \mathbf{u})) = \nabla\mu\nabla \cdot \mathbf{u} + \mu\nabla^2\mathbf{u} \qquad (7.4)$$

where the terms in the right-hand side of Eq. (7.4) can be written as

$$\langle\nabla\mu\nabla \cdot \mathbf{u}\rangle_i = \left(\frac{d}{n^0}\right)^2 \sum_{j\neq i}\left(\frac{\mathbf{u}_j - \mathbf{u}_i}{r_{ij}} \cdot \mathbf{e}_{ij}w\left(r_{ij}, r_e\right)\right) \sum_{j\neq i}\left(\frac{\mu_j - \mu_i}{r_{ij}} \cdot \mathbf{e}_{ij}w\left(r_{ij}, r_e\right)\right)$$

$$(7.5)$$

$$\langle\mu\nabla^2\mathbf{u}\rangle_i = \frac{2d}{\lambda n^0} \sum_{j\neq i}\left(\mu_{ij}\left(\mathbf{u}_j - \mathbf{u}_i\right) \cdot \mathbf{e}_{ij}w\left(r_{ij}, r_e\right)\right) \qquad (7.6)$$

where $\mathbf{e}_{ij}$ is the unit vector; μ_{ij} is the interaction viscosity between particles i and j. When particles i and j belong to the same phase, the actual viscosity is applied, that is, $\mu_{ij} = \mu_i = \mu_j$. When particles i and j belong to different phases, none of their actual viscosities can be used. Therefore, a multiviscosity fluid is modeled on the basis of dissipative particle dynamics, and the interaction viscosity is calculated by

$$\mu_{ij} = \left(\left(\mu_i^\theta + \mu_j^\theta\right)/2\right)^{\frac{1}{\theta}} \qquad (7.7)$$

where θ is an adjusting parameter. Eq. (7.7) returns the arithmetic mean for $\theta = 1$ and harmonic mean for $\theta = -1$. Application of the interaction viscosity in calculation of the viscous term automatically adds the multiphase forces (arising from the viscosity difference) to the momentum equation [5].

Khayyer and Gotoh [6] smoothed the density by the first-order Taylor series expansion to model an accurate and consistent phase interface:

$$\rho_i = \frac{1}{\sum_{j\in I}w_{ij}} \sum_{j\in I}\left(\rho_j - \frac{\partial\rho_i}{\partial x_{ij}}x_{ij} - \frac{\partial\rho_i}{\partial y_{ij}}y_{ij}\right)w_{ij} \qquad (7.8)$$

where I corresponds to target particle i and all its neighboring particles. Compared with the density smoothing by Eq. (7.1) that has a zeroth-order accuracy, Eq. (7.8) has the first-order accuracy. Fig. 7.4 shows the density variations at the interface using the density smoothing schemes of Eq. (7.1) and Eq. (7.8). It can be seen that the Taylor series expansion density smoothing gives a more accurate evaluation of density at target particle i and provides a sharp interface. It is because the density smoothing scheme derived from Taylor series expansion considers the spatial distributions of densities at neighboring particles and the spatial variations of density at the target particle itself. The interface sharpness compared between the spatial average density smoothing (Eq. 7.1) and Taylor series expansion density smoothing (Eq. 7.8) is shown in Fig. 7.5. The Taylor series expansion density smoothing is proved to be effective in preserving the sharpness of spatial density variations and provides a continuous pressure field by minimizing unphysical perturbations.

Duan et al. [7] analyzed the pressure gradient force and acceleration velocity in the interaction of light and heavy particles and pointed out

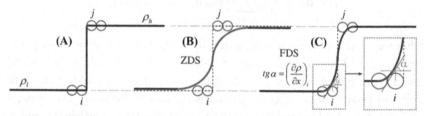

Figure 7.4 Density variation at the interface, (A) without density smoothing, (B) zeroth-order density smoothing (ZDS) by Eq. (7.1), and (C) first-order density smoothing (FDS) by Eq. (7.8) [6].

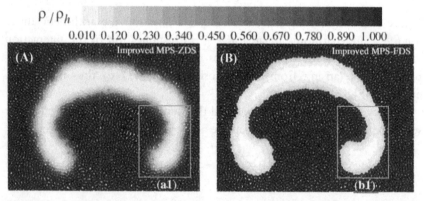

Figure 7.5 Interface sharpness with density smoothed by (A) Eq. (7.1) and (B) Eq. (7.8) [6].

that the large acceleration of light particles is the cause of instability. Thus, they pay their attention to the density smoothing in the pressure Poisson equation (PPE) and gradient model improvement. In the calculation of gas–liquid two-phase flow, the PPE source term for both phases is the same, but the matrix coefficients for pressure calculation have a large difference because the coefficients in the matrixes are related to the density. The coefficients in the matrix of the gas phase are very large, resulting in a small pressure variation, but it is opposite in the liquid phase. Duan et al. [7] tested the effect of different PPE density average schemes on the interface pressure variation. Fig. 7.6A indicates that the arithmetic average of gas and liquid densities can achieve an accurate pressure profile, which matches well with the real values. However, this pressure profile is not good to the stability. It is because for gas particle i, its neighbor liquid particle j has a large pressure, which results in a large acceleration velocity for gas particle i. Comparatively, the harmonic density average in PPE can move the pressure transition line to the liquid phase side, as shown in Fig. 7.6B. In this condition, the pressure at the interface closes to that of the gas phase, and thus the acceleration velocity of the gas particle is not extremely large. However, the pressure profile at the interface does not match with the real pressure, and the calculation accuracy is sacrificed.

On the basis of accurate pressure calculation with arithmetic density average in PPE, Duan et al. [7] analyzed acceleration velocities that are induced by pressure gradient force at gas and liquid particles. As shown in Fig. 7.7, the pressure gradient force is the same for particles i and j that have a density ratio of 1000. If the force directly exerts on particles i and j, their acceleration velocities have a large difference because the force is divided by their respective densities. To solve this problem, therefore,

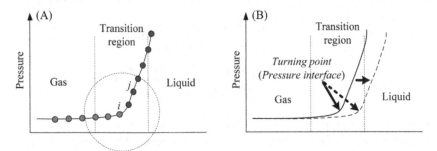

Figure 7.6 Pressure profiles near the gas–liquid two-phase interface, (A) arithmetic density average in PPE and (B) arithmetic density average (solid line) and the harmonic density average (dashed line) in PPE [7].

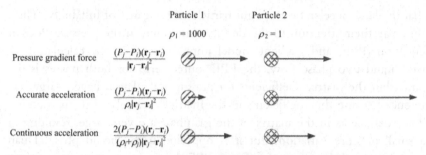

Figure 7.7 Sketch for the acceleration with different discretizing formulations [7].

Duan et al. [7] used arithmetic density average in the pressure gradient model, which results in a continuous acceleration of particles i and j. Similarly, in the solving of viscosity and surface tension models where density presents, the arithmetic density average is adopted as well.

It should be noted that if we directly use either of the following two pressure gradient models even with the arithmetic density average being adopted (Eqs. 7.9 and 7.10), the calculation is not stable.

$$\left\langle \frac{1}{\rho} \nabla P \right\rangle_i = \frac{d}{n^0} \sum_{j \neq i} \frac{2(P_j - P_i)(\mathbf{r}_j - \mathbf{r}_i)}{(\rho_i + \rho_j)|\mathbf{r}_j - \mathbf{r}_i|^2} \mathbf{C}_{ij} w_{ij} \qquad (7.9)$$

$$\left\langle \frac{1}{\rho} \nabla P \right\rangle_i = \frac{d}{n^0} \sum_{j \neq i} \frac{2(P_j - P_{i,\min})(\mathbf{r}_j - \mathbf{r}_i)}{(\rho_i + \rho_j)|\mathbf{r}_j - \mathbf{r}_i|^2} \mathbf{C}_{ij} w_{ij} \qquad (7.10)$$

It is because Eq. (7.9) does not use the term $P_{i,\min}$, and as a result, attractive force presents in the particle interactions. Even though Eq. (7.10) converts attractive force to repulsive force, the instability presents as well. In order to find the reasons, Eq. (7.10) is rewritten as [7]

$$\left\langle \frac{1}{\rho} \nabla P \right\rangle_i = \frac{d}{n^0} \sum_{j \neq i} \frac{2(P_j - P_i)(\mathbf{r}_j - \mathbf{r}_i)}{(\rho_i + \rho_j)|\mathbf{r}_j - \mathbf{r}_i|^2} \mathbf{C}_{ij} w_{ij}$$
$$+ \frac{d}{n^0} \sum_{j \neq i} \frac{2(P_i - P_{i,\min})(\mathbf{r}_j - \mathbf{r}_i)}{(\rho_i + \rho_j)|\mathbf{r}_j - \mathbf{r}_i|^2} \mathbf{C}_{ij} w_{ij} \qquad (7.11)$$

where the first term at the right-hand side is the same as Eq. (7.9) and the second term is defined as the pressure stability term. The pressure stability term is not constant for particle i because the density of particle j presents in the denominator. It is the cause of instability. Specifically speaking, if particles i and j are both gas particles, the value of $2(P_j - P_{i,\min})/(\rho_i + \rho_j)$

is large; otherwise, if particles i and j are respectively gas and liquid particles, the value of $2(P_j - P_{i,\min})/(\rho_i + \rho_j)$ is small. It is equivalent to that gas particle j has a high pressure, while liquid particle j has a low pressure. As a result, the pressure stability term of Eq. (7.11) will drive gas particles to move to the liquid particles, which leads to calculation instability. Duan et al. [7] proposed to use only the density of particle i in the pressure stability term of Eq. (7.11), and the equation of pressure gradient calculation is written as

$$
\left\langle \frac{1}{\rho}\nabla P \right\rangle_i = \frac{d}{n^0}\sum_{j\neq i}\frac{2(P_j - P_i)(\mathbf{r}_j - \mathbf{r}_i)}{(\rho_i + \rho_j)|\mathbf{r}_j - \mathbf{r}_i|^2}\mathbf{C}_{ij}w_{ij}
$$
$$
+ \frac{d}{n^0}\sum_{j\neq i}\frac{(P_i - P_{i,\min})(\mathbf{r}_j - \mathbf{r}_i)}{\rho_i|\mathbf{r}_j - \mathbf{r}_i|^2}\mathbf{C}_{ij}w_{ij}
$$

(7.12)

It should be noted that the value of $P_{i,\min}$ is another trigger of instability. For instance, if the $P_{i,\min}$ of gas particle i presents at a liquid particle j, the value of $(P_i - P_{i,\min})/\rho_i$ may be larger than the value of $2(P_j - P_i)/(\rho_i + \rho_j)$ in the real pressure gradient term because ρ_i is much smaller than ρ_j. In such conditions, the pressure stability term will exert overadjustment compared with the real pressure gradient term. Therefore, the $P_{i,\min}$ is defined to be searched only in the same phase of particle i. The final pressure gradient equation of gas–liquid two-phase flow is written as

$$
\left\langle \frac{1}{\rho}\nabla P \right\rangle_i = \frac{d}{n^0}\sum_{j\neq i}\frac{2(P_j - P_i)(\mathbf{r}_j - \mathbf{r}_i)}{(\rho_i + \rho_j)|\mathbf{r}_j - \mathbf{r}_i|^2}\mathbf{C}_{ij}w_{ij}
$$
$$
+ \frac{d}{n^0}\sum_{j\neq i}\frac{(P_i - P'_{i,\min})(\mathbf{r}_j - \mathbf{r}_i)}{\rho_i|\mathbf{r}_j - \mathbf{r}_i|^2}\mathbf{C}_{ij}w_{ij}
$$

(7.13)

Fig. 7.8 shows the simulation of a bubble rising in the liquid with the pressure gradient model of Eq. (7.13), in which the density and viscosity ratios of liquid and gas are 1000 and 100, respectively. It can be seen that the simulation is stable and the bubble shape can match with the benchmark case.

Moreover, some artificial interparticle forces are applied to the simulation of multiphase flow with large density and viscosity differences. Wang and Zhang [9] applied the peridynamic theory to MPS method to study multiphase flow with discontinuity. The stress tensor at the fluid interface is solved by MPS method. The integral form of the governing equation can reduce the influence of density and viscosity discontinuities on the

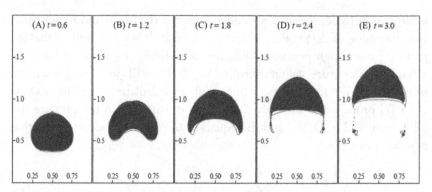

Figure 7.8 Simulation of a bubble rising in the liquid with the pressure gradient model of Eq. (7.13) (the line represents the results of benchmark case by Hysing et al. [8]) [7].

accuracy and stability. In addition, the divergence decomposition of stress tensor and harmonic mean density are adopted to reduce the influence of density discontinuity on interface calculation. The pressure and the velocity gradient are calculated based on the explicit scheme and the particle interaction models, respectively. A modified matrix is also added to the decomposition equation to deal with the irregular particle distribution.

7.2 Solid—liquid two-phase flow

The methods for numerical simulation of solid—liquid flows can be classified as the discrete description and the continuum description of solid granular matter. In the discrete description method, the macroscopic and microscopic forces acting on the individual granules are calculated. In such kinds of methods, the discrete element method (DEM) is widely used in the simulation of granular flows, which evaluates the contact forces based on Newton's second law of motion. With the combination of DEM and MPS, the solid—liquid flow can be effectively simulated, in which the DEM calculates the motion of submerged granules, while the MPS method calculates the continuum liquid flow. In contrast to the discrete description of granular flow, the continuum description of granular flow treats the solid body as a complex fluid (e.g., non-Newtonian fluid) and solves the conservation equations of mass and momentum to predict the state of the flow system. The contact forces of solid granules are not solved. The numerical models of discrete description and continuum description of solid granules are introduced in the following parts.

7.2.1 Discrete description of solid granules

Sakai et al. [10] developed an MPS-DEM method to simulate solid–liquid flows involving the free surface, in which the motion of the liquid phase is modeled with MPS method and the solid phase is solved using DEM. The solid and liquid interaction forces including drag force, contact force, gravitational force, virtual mass force, lubrication force, and pressure force are solved. Taking these forces into consideration the equation of state of the solid phase is written as

$$m_s \frac{d\mathbf{v}_s}{dt} = \mathbf{F}_d - V_s \nabla p + \sum \mathbf{F}_c + \mathbf{F}_g + \mathbf{F}_{vm} + \mathbf{F}_l \qquad (7.14)$$

where m, $\mathbf{v}$, and V are the particle mass, particle velocity, and particle volume, respectively, and the subscript s denotes the solid particles; moreover, $\mathbf{F}_d$, $\mathbf{F}_c$, $\mathbf{F}_g$, $\mathbf{F}_{vm}$, $\mathbf{F}_l$, and p are the drag force, contact force, gravitational force, virtual mass force, lubrication force, and fluid pressure, respectively. The following calculation details of each force come from the Ref. [10].

For rotational motion, the angular acceleration of solid particles is calculated by

$$\frac{d\boldsymbol{\omega}_s}{dt} = \frac{\mathbf{T}}{I} \qquad (7.15)$$

where $\boldsymbol{\omega}$ is the angular velocity, I is the moment of inertia, and $\mathbf{T}$ is the interparticle contact torque.

The contact force acting on a solid particle is calculated from the overlap magnitude of particle–particle or particle–wall by using a spring, dashpot, and friction slider. It is composed of normal and tangential components. The normal component of the contact force is calculated by

$$\mathbf{F}_{c_n} = -k\boldsymbol{\delta}_n - \eta \mathbf{v}_{s_n} \qquad (7.16)$$

where k, $\boldsymbol{\delta}$, and η are the spring, displacement, and damping coefficients, respectively, and the subscript n denotes the normal component. The tangential component of the contact force is given by

$$\mathbf{F}_{c_t} = \begin{cases} -k\boldsymbol{\delta}_t - \eta \mathbf{v}_{s_t} & \left(\left| \mathbf{F}_{c_t} \right| < \mu \left| \mathbf{F}_{c_n} \right| \right) \\ -\mu \left| \mathbf{F}_{c_n} \right| \mathbf{v}_{s_t} / \left| \mathbf{v}_{s_t} \right| & \left(\left| \mathbf{F}_{c_t} \right| \geq \mu \left| \mathbf{F}_{c_n} \right| \right) \end{cases} \qquad (7.17)$$

where μ is the friction coefficient and the subscript t represents the tangential component. η satisfies the following equations:

$$\eta = - 2\ln(e)\left(\sqrt{\frac{km_s}{\ln^2(e) + \pi^2}}\right) \qquad (7.18)$$

where e is the restitution coefficient.

The calculation of drag force depends on not only the relative velocity between the solid particle and liquid fluid but also the presence of neighboring particles, that is, local volume fraction of the solid phase. Therefore, the equation for drag force calculation is expressed as

$$\mathbf{F}_d = \frac{\beta}{1 - \varepsilon}\left(\mathbf{u}_f - \mathbf{v}_s\right) V_s \qquad (7.19)$$

where ε, $\mathbf{u}_f$, and β are the fluid volume fraction, fluid velocity, and interphase momentum transfer coefficient, respectively. A volume fraction of $\varepsilon = 0.8$ is used as the boundary of a piecewise function for the calculation of interphase momentum transfer coefficient β:

$$\beta = \begin{cases} 150\dfrac{(1-\varepsilon)^2}{\varepsilon}\dfrac{\mu_f}{d_s^2} + 1.75(1-\varepsilon)\dfrac{\rho_f}{d_s}\left|\mathbf{u}_f - \mathbf{v}_s\right| & (\varepsilon \leq 0.8) \\[2ex] \dfrac{3}{4}C_d\dfrac{\varepsilon(1-\varepsilon)}{d_s}\rho_f\left|\mathbf{u}_f - \mathbf{v}_s\right|\varepsilon^{-2.65} & (\varepsilon > 0.8) \end{cases} \qquad (7.20)$$

where μ_f, d_s, ρ_f, and C_d are the fluid viscosity, solid particle diameter, fluid density, and drag coefficient for an isolated particle, respectively. The drag coefficient C_d is determined by

$$C_d = \begin{cases} \dfrac{24}{Re_s}\left(1 + 0.15Re_s^{0.687}\right) & (Re_s \leq 1000) \\[2ex] 0.44 & (Re_s > 1000) \end{cases} \qquad (7.21)$$

where the Reynolds number (Re_s) is

$$Re_s = \frac{\left|\mathbf{u}_f - \mathbf{v}_s\right|\varepsilon\rho_f d_s}{\mu_f} \qquad (7.22)$$

The lubrication force is derived from the hydrodynamic pressure when the interstitial fluid is squeezed out of the space between two solid particles. The equation of the lubrication force is given by

$$\mathbf{F}_l = \frac{3\pi\mu_f d_s^2 \left(\mathbf{v}_{s_j} - \mathbf{v}_{s_i}\right)}{8\left(\left|\mathbf{r}_j - \mathbf{r}_i\right| - d_s\right)} \tag{7.23}$$

The virtual mass force is related to the work corresponding to the acceleration of solid versus fluid. The equation of virtual mass force is expressed as follows:

$$\mathbf{F}_{vm} = \frac{1}{2}\rho_f V_s \left(\frac{D\mathbf{u}_f}{Dt} - \frac{d\mathbf{v}_s}{dt}\right) \tag{7.24}$$

The solid particle acceleration is calculated by considering the solid–solid interaction force, solid–fluid interaction force, and external force. The angular acceleration is obtained from torques. The equations of particle velocity $\mathbf{v}_s$, particle position $\mathbf{x}_s$, angular velocity $\boldsymbol{\omega}_s$, and angle $\boldsymbol{\theta}_s$ are written as

$$\mathbf{v}_s^{n+1} = \mathbf{v}_s^n + \frac{d\mathbf{v}_s^n}{dt}\Delta t \tag{7.25}$$

$$\mathbf{x}_s^{n+1} = \mathbf{x}_s^n + \mathbf{v}_s^{n+1}\Delta t \tag{7.26}$$

$$\boldsymbol{\omega}_s^{n+1} = \boldsymbol{\omega}_s^n + \frac{d\boldsymbol{\omega}_s^n}{dt}\Delta t \tag{7.27}$$

$$\boldsymbol{\theta}_s^{n+1} = \boldsymbol{\theta}_s^n + \boldsymbol{\omega}_s^{n+1}\Delta t \tag{7.28}$$

where $\mathbf{x}$ and $\boldsymbol{\theta}$ are the position vector of center of mass and angular displacement, respectively, and the superscript n denotes time step.

In the MPS-DEM method the continuity and momentum equations of the liquid phase are modified by considering the local average treatment.

$$\frac{D\hat{\rho}_f}{Dt} + \hat{\rho}_f \nabla \cdot \mathbf{u}_f = 0 \tag{7.29}$$

$$\hat{\rho}_f \frac{D\mathbf{u}_f}{Dt} = -\varepsilon\nabla P + \mathbf{f} + \varepsilon\nabla \cdot \boldsymbol{\tau}_f + \hat{\rho}_f \mathbf{g} \tag{7.30}$$

where $\hat{\rho}_f = \varepsilon\rho_f$, and $\boldsymbol{\tau}_f$, $\mathbf{f}$, and $\mathbf{g}$ are the molecular viscous stress tensor, solid–liquid interaction body force, and gravitational acceleration, respectively. The solid–liquid interaction body force $\mathbf{f}$ is a volume force calculated by

$$\mathbf{f} = \frac{\sum_{i=1}^{N} \mathbf{F}_{d_i}}{V_e} \tag{7.31}$$

where V_e is the volume confined by effective radius r_e, N is the number of solid particles in the effective volume, and $\mathbf{F}_{d_i}$ is the drag force of particle i calculated by Eq. (7.19).

The modeling procedures of the liquid phase are the same as that of the original MPS method, except that the volume fraction of the solid phase should be taken into consideration when calculating the PND. For the incompressible fluid, the PND of each particle is maintained to be the average PND, that is,

$$\hat{n}_i = \varepsilon n^0 = \hat{n}^0 \tag{7.32}$$

Although the DEM provides accurate calculations of solid—solid and solid—liquid interaction forces, the calculation cost is too expansive because of the large number of solid and liquid particles. Sakai and Koshizuka [11] and Sakai et al. [12] have proposed a coarse grain model, which uses a large-sized particle (or named coarse grain particle) to represent a group of original particles, as illustrated in Fig. 7.9. One coarse grain particle contains l^3 original particles, where l means the scale ratio of

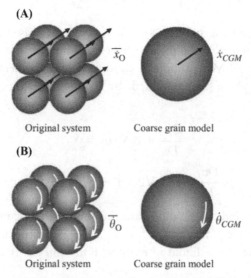

Figure 7.9 A coarse grain model, (A) translational motion and (B) rotational motion [11].

coarse grain particle to the original particle. In coarse grain model the contact force is modeled under the assumption that the kinetic energy of the coarse grain particle agrees with that of original particles. The drag force and the external forces such as gravitational force are modeled by balancing the coarse grain particle with the group of the original particles. The van der Walls force is modeled based on the assumption that the potential energy of the coarse grain particle is the same as that of the original particles. This modeling approach is different from that one directly uses a large-sized solid particle.

The conservation of kinetic energy between the coarse grain particle and the group of original particles is given by

$$\frac{1}{2} m_{CGM} \mathbf{v}^2_{CGM} + \frac{1}{2} I_{CGM} \boldsymbol{\omega}^2_{CGM} = l^3 \left(\frac{1}{2} m_O \overline{\mathbf{v}}^2_O + \frac{1}{2} I_O \overline{\boldsymbol{\omega}}^2_O \right) \qquad (7.33)$$

where the subscripts of CGM and O denote the coarse grain particle and the original particle, respectively, and the upper line stands for the average value.

As illustrated by Fig. 7.9, for the translational motion the velocity of the coarse grain particle is equal to the average velocity of the group of original particles, while for the rotational motion the original particles included in the coarse grain particle are assumed to rotate around their own center of mass with an equal angular velocity. The rotation of original particles around the center of mass of the coarse grain particle is not considered.

In the rotational motion, the relation of angular acceleration between the original particles and the coarse grain particle is expressed as

$$\frac{d\boldsymbol{\omega}_{CGM}}{dt} = \frac{\mathbf{T}_{CGM}}{I_{CGM}} = \frac{\mathbf{r}_{CGM} \times \mathbf{F}_{CGM_t}}{I_{CGM}} = \frac{l\mathbf{r}_O \times l^3 \mathbf{F}_{O_t}}{l^5 I_O} = \frac{l^4 \overline{\mathbf{T}}_O}{l^5 I_O} = \frac{1}{l} \frac{d\overline{\boldsymbol{\omega}}_O}{dt}$$
$$(7.34)$$

The contact force at the normal direction when binary coarse grain particles i and j collide is written as follows:

$$\mathbf{F}_{c_{CGM_n}} = l^3 \overline{\mathbf{F}}_{c_{O_n}} = l^3 \left(-k \overline{\boldsymbol{\delta}}_{O-iaja_n} - \eta \overline{\mathbf{v}}_{O-iaja_n} \right) = l^3 \left(-k \boldsymbol{\delta}_{CGM-ij_n} - \eta \mathbf{v}_{CGM-ij_n} \right)$$
$$(7.35)$$

where the subscripts i a and j a represent an arbitrary pair of original particles in binary collision of coarse grain particles i and j. The translational motion of coarse grain particles can be assumed to agree with that of the original particles, and thus Eq. (7.35) is rewritten as

$$F_{c_{CGM_n}} = l^3 \left(-k\delta_{CGM-ij_n} - \eta v_{CGM-ij_n} \right) \tag{7.36}$$

For the coarse grain particles that have no slippage at contact point $\left(\left| F_{c_{CGM_t}} \right| < \mu \left| F_{c_{CGM_n}} \right| \right)$, the tangential component of the contact force is given by

$$F_{c_{CGM_t}} = l^3 \overline{F}_{c_{O_t}} = l^3 \left(-k\overline{\delta}_{O-iaja_t} - \eta\overline{v}_{O-iaja_t} \right) = l^3 \left(-k\delta_{CGM-ij_t} - \eta v_{CGM-ij_t} \right) \tag{7.37}$$

When the slippage occurs at the contact point of two coarse grain particles $\left(\left| F_{c_{CGM_t}} \right| \geq \mu \left| F_{c_{CGM_n}} \right| \right)$, the tangential component of the contact force is modified as

$$\begin{aligned} F_{c_{CGM_t}} &= - \mu l^3 \left| \overline{F}_{O_n} \right| \overline{v}_{O_t} / \left| \overline{v}_{O_t} \right| \\ &= - \mu \left| F_{CGM_n} \right| v_{CGM_t} / \left| v_{CGM_t} \right| \end{aligned} \tag{7.38}$$

The number of contact points in coarse grain model is determined by multiplying the calculated number of contact points with l^3. As a result the particle–particle interactions of all the original particles that are included in the coarse grain particles are taken into consideration.

The drag force and the external force are modeled by balancing a coarse grain particle and a group of original particles. The drag force acting on a coarse grain particle is calculated by

$$F_{d_{CGM}} = \frac{\beta}{1 - \varepsilon} \left(u_f - v_{CGM} \right) V_{CGM} = l^3 \frac{\beta}{1 - \varepsilon} \left(u_f - \overline{v}_O \right) V_O \tag{7.39}$$

The van der Waals force is modeled by assuming that the potential energy of the coarse grain particle is equal to that of the group of original particles. Hence, the equation of the van der Waals force acting on the coarse grain particle is calculated as

$$F_{vdw_{CGM}} = l^2 \overline{F}_{vdw_O} = l^3 \frac{H_A d_{s_{CGM}}}{6h_{CGM}^2} n = l^2 \frac{H_A d_{s_O}}{6h_O^2} n \tag{7.40}$$

where H_A and h are the Hamaker constant and intersurface distance, respectively. In Eq. (7.40), $h_O = h_{CGM}/l$.

Eventually, the equation of state of the coarse grain particle, which takes into consideration of the contact force, the drag force, the gravitational force, and the van der Waals force, is written as

$$m_{CGM}\frac{d\mathbf{v}_{CGM}}{dt} = \mathbf{F}_{d_{CGM}} - V_{CGM}\nabla p + \sum \mathbf{F}_{c_{CGM}} + \mathbf{F}_{g_{CGM}} + \mathbf{F}_{vdw_{CGM}}$$
$$= l^3\overline{\mathbf{F}}_{d_O} - l^3 V_O\nabla p + l^3 \sum \overline{\mathbf{F}}_{c_O} + l^3\overline{\mathbf{F}}_{g_O} + l^2\overline{\mathbf{F}}_{vdw_O} \tag{7.41}$$

The mass and momentum conservation equations of the liquid phase are the same as Eqs. (7.29) and (7.30).

Modeling the solid–liquid flow with the above coarse grain particle model can effectively reduce the computation cost and memory resource. It makes the MPS-DEM method capable of simulating the large-scale flows.

7.2.2 Continuum description of the solid granules

There is another multiphase modeling method which treats the fluids as a single-phase, multidensity, and multiviscosity system, and only a set of hydrodynamic equations need to be solved. This method is defined as a continuum description of the solid granules. The interaction forces of solid particles are not modeled.

Regarding the continuum description of solid granules in MPS method, Koshizuka et al. [13] first proposed a passively moving solid model to calculate the motion of solid bodies in fluids. The solid body is represented by a cluster of particles which have a fixed relative configuration. The center coordinates $\mathbf{r}_g$ of the solid body are evaluated from the particle coordinates $\mathbf{r}_i$:

$$\mathbf{r}_g = \frac{1}{N}\sum_{i=1}^{N} \mathbf{r}_i \tag{7.42}$$

In the calculation algorithm, the solid particles are first treated to be the same as fluid particles, and the pressure and velocity of all the particles are simultaneously calculated using the hydrodynamic equations. It should be noted that the solid particles are not moved with their individual velocities. Instead, an additional procedure is applied to the solid particles to calculate the transition and rotation velocity vectors of the solid body using Eqs. (7.43) and (7.44). After that, the average velocity of solid particles is inversely calculated from the translation and rotation velocity vectors and the relative coordinates of solid particles by Eq. (7.45). The original velocities of solid particles are replaced with the average velocity.

This passively moving solid model only considers the hydraulic forces but does not consider the solid contact forces. The fast calculation is its advantage, but the accuracy is sacrificed.

$$\mathbf{T} = \frac{1}{N} \sum_{i=1}^{N} \mathbf{u}_i \tag{7.43}$$

$$\mathbf{R} = \frac{1}{I} \sum_{i=1}^{N} \mathbf{u}_i \times \mathbf{q}_i \tag{7.44}$$

$$\mathbf{u}_i = \mathbf{T} + \mathbf{q}_i \times \mathbf{R} \tag{7.45}$$

where $\mathbf{T}$ and $\mathbf{R}$ are the translation and rotation velocity vectors, respectively, I is the moment of inertia, $\mathbf{q}$ is the relative coordinates of solid particles, and N is the particle number in one solid body.

Different from Koshizuka's passively moving solid model, there is another treatment which imagines the mixture of solid granules and pore fluids as non-Newtonian fluids, while the ambient fluid is a Newtonian fluid, as shown in Fig. 7.10. The Newtonian and non-Newtonian fluids are calculated with different types of particles. Different from the Newtonian fluid, the viscosity of the non-Newtonian fluid cannot be determined simply by Newtonian model. Hence, the rheology model is widely used to determine the viscosity of non-Newtonian fluids, that is, the mixture of the solid granules and pore fluids. The transition viscosity at the phase interface of Newtonian and non-Newtonian fluids is usually calculated by harmonic average.

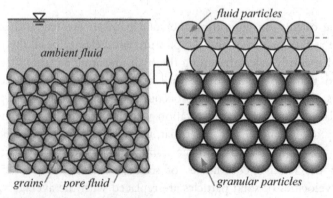

Figure 7.10 Particle representation [14].

The rheology of non-Newtonian fluids can be expressed by a nonlinear constitutive relation between shear stress and strain tensor:

$$\boldsymbol{\tau} = F(\mathbf{E}) = \phi_0 \mathbf{I} + \phi_1 \mathbf{E} + \phi_2 \mathbf{E}^2 \tag{7.46}$$

where $\boldsymbol{\tau}$ is the shear stress tensor, $\mathbf{E}$ is the rate of strain tensor, and ϕ is the scalar function of three principal invariants of tensor $\mathbf{E}$, namely, I_E, II_E, and III_E (the first, second, and third invariants, respectively). By including $\phi_0 \mathbf{I}$ into the thermodynamic pressure term $-P\mathbf{I}$ and assuming a constant density ($I_E = 0$), the equation is simplified as

$$\boldsymbol{\tau} = \phi_1(II_E, III_E)\mathbf{E} + \phi_2(II_E, III_E)\mathbf{E}^2 \tag{7.47}$$

which is the general form of constitutive equation for defining a Reiner–Rivlin fluid.

There are several rheology models to calculate the shear stress and the effective viscosity of non-Newtonian fluids. In the following part, the Bingham plastic model, the Herschel–Bulkley model, and the generalized viscoplastic fluid model are introduced.

The Bingham plastic model is a two-parameter rheology model assuming that no deformation takes place until the stress beyond the yield stress τ_y. The shear stress in the Bingham plastic model is calculated by

$$\boldsymbol{\tau} = \begin{cases} \left(\dfrac{\tau_y}{\sqrt{II_E}} + 2\mu_0 \right) \mathbf{E} & |\boldsymbol{\tau}| > \tau_y \\ 0 & |\boldsymbol{\tau}| \le \tau_y \end{cases} \tag{7.48}$$

Therefore, the effective viscosity μ_{eff} of the viscous non-Newtonian fluid is calculated by

$$\mu_{\text{eff}} = \frac{\tau_y}{\sqrt{4II_E}} + \mu_0 \tag{7.49}$$

where μ_0 is the flow consistency index.

The Herschel–Bulkley model is a generalized viscoplastic model for the non-Newtonian fluid, which combines the power-law model and Bingham plastic model. The shear stress in the Herschel–Bulkley model is calculated by

$$\boldsymbol{\tau} = \begin{cases} \left(\dfrac{\tau_y}{\sqrt{II_E}} + 2\mu_0 \left(\sqrt{4II_E}\right)^{N-1} \right) \mathbf{E} & |\boldsymbol{\tau}| > \tau_y \\ 0 & |\boldsymbol{\tau}| \le \tau_y \end{cases} \tag{7.50}$$

where N is the flow behavior index. The effective viscosity when the stress is higher than the yield stress is given as

$$\mu_{eff} = \frac{\tau_y}{\sqrt{4II_E}} + \mu_0\left(\sqrt{4II_E}\right)^{N-1} \tag{7.51}$$

The generalized viscoplastic fluid model (GVF) is a semiempirical formula proposed by Chen et al. [15] for the granular flow. The shear stress and the effective viscosity in this model are calculated by

$$\boldsymbol{\tau} = \left(\frac{\tau_y}{\sqrt{II_{E'}}} + 2\mu_1\left(\sqrt{4II_{E'}}\right)^{N_1-1}\right)\mathbf{E} + \left(4\mu_2\left(\sqrt{4II_{E'}}\right)^{N_2-1}\right)\mathbf{E}^2 \tag{7.52}$$

$$\mu_{eff} = \left(\frac{\tau_y}{\sqrt{4II_{E'}}} + \mu_1\left(\sqrt{4II_{E'}}\right)^{N_1-1}\right) + \left(2\mu_2\left(\sqrt{4II_{E'}}\right)^{N_2-1}\right) \tag{7.53}$$

where μ_1 and μ_2 are the flow behavior indexes, and N_1 and N_2 are the consistency and cross consistency indexes, respectively. $\mathbf{E}'$ is the deviation rate of strain tensor, and it is calculated by

$$\mathbf{E}' = \mathbf{E} - \frac{1}{3}(\mathrm{tr}\mathbf{E})\mathbf{I} \tag{7.54}$$

For isotropic granular materials, the expression of yield criteria is defined as

$$\tau_y = c\cos\varphi + \bar{p}\sin\varphi \tag{7.55}$$

where c, φ, and $\bar{p}$ are the cohesion coefficient, internal friction angle, and hydrostatic pressure, respectively. When ignoring the second term of GVF model, it will be simplified to the Herschel–Bulkley model.

With the above models, the simulations of granular flows have been widely conducted. Nodoushan et al. [14] modeled multiphase flow of granular materials and fluids by treating it as a multidensity and multiviscosity system, and a single set of governing equations is solved for the entire flow field. The effective viscosity of the granular phase is evaluated using a regularized Herschel–Bulkley rheological model with a pressure-dependent yield criterion. The weakly compressible scheme is applied for the stability of the calculation. Many adjusting coefficients are adopted in the calculation models of effective viscosity and pressure, and they are calibrated in the simulations of dry and submerged granular flows. Tajnesaie et al. [16] replaced the Herschel–Bulkley rheological model in Nodoushan's modeling by a new viscoplastic rheological model, which is a function of dynamic intergrain mechanical pressure. The results present more accurate granular fields. Jandaghian et al. [17] developed a single-phase particle method to simulate the immersed dense granular flows based on the enhanced weakly compressible MPS method. The regularized viscoinertial model, consistent effect, and higher-order operators are employed in rheology model. The turbulent viscosity and linear effects of interfacial suspension are considered within the constitutive law.

7.3 Turbulence model

Even though the MPS method has been used to simulate various violent free-surface flows, the turbulence model is usually not applied. The Froude number usually plays a more important role in most free-surface flows compared to the Reynolds number. In other words, the diffusion due to the molecular viscosity or the eddy viscosity may be secondary for many flow characteristics. Thus, the diffusion term and the turbulence effect can be neglected. Nevertheless, it is worth to consider the turbulent effect in many flows, especially when the diffusion terms (e.g., heat transfer and mass transfer) are considered.

The turbulence models can be briefly clarified into the Reynolds-averaged Navier–Stokes equations method (RANS) and the large eddy simulation (LES). Considering that LES can provide more turbulent information than RANS, the LES-based particle method is presented in this section. Like the subgrid scale (SGS) models in mesh-based LES, subparticle scale (SPS) models are necessary to represent the effect of unresolved subparticle scale on the resolved particle scale flow. Lo and Shao [18] first developed the static Smagorinsky model for the particle method. Similar SPS models are also adopted in particle method to simulate wave-breaking flow [19], dam-breaking flow [19], plunging wave flow [20], and wave overtopping-deck flow [21]. Arai et al. [22] developed a turbulence wall model to simulate the flow around a cubic obstacle and compared the velocity fluctuations with experimental results. Duan and Chen [23] developed a dynamic Smagorinsky model and simulated the typical turbulent flow, namely, the mixing layer. The vortex pair phenomena were well reproduced, and many turbulent characteristics were compared with the experimental measurements.

Nevertheless, the turbulence models in particle methods face more difficulties than those in the mesh methods. Specifically, particle methods are less accurate than mesh methods. Thus, numerical fluctuations can unavoidably take place in the simulations by particle methods. The numerical fluctuations can obscure the real turbulent fluctuations to some extent, as pointed out in Refs. [22,23]. This can make the turbulence models less effective in particle methods, which is the main challenge. From this viewpoint, developing accurate and convergent particle methods can help to improve the effectiveness of turbulence models.

On the other hand, the turbulence models are usually employed in particle methods for two purposes. First, the turbulence models can increase the viscosity dissipation. When fine particles are used, the viscous dissipation may be underestimated due to the lack of turbulence effects. In this situation, the turbulence model can make the simulation more reasonable. When the rather violent flow is simulated even by coarse particles, the consideration of turbulence models can greatly enhance the stability. Second, the turbulence models can improve the heat transfer. When the heat transfer problem is simulated by large particles, the convective heat transfer may be significantly underestimated. In this situation, the turbulence models can help to produce more reasonable results.

The typical turbulence models employed in particle methods are presented below. In large eddy simulations, a spatial filter is applied to each variable to separate the resolved particle scale portion and the unresolved subparticle scale portion as follows:

$$\phi = \overline{\phi} + \phi' \tag{7.56}$$

where ϕ is an arbitrary variable, $\overline{\phi}$ is the resolved particle scale component, and ϕ' is the unresolved subparticle portion. The contribution of the subparticle scales to the particle scales is modeled. The filtered Lagrangian equations are as follows:

$$\frac{D\overline{\rho}}{Dt} \approx 0 \tag{7.57}$$

$$\frac{D\overline{\mathbf{u}}}{Dt} = -\frac{1}{\rho}\nabla\overline{p} + \nu\nabla^2\overline{\mathbf{u}} + \overline{\mathbf{g}} + \nabla\cdot\boldsymbol{\tau} \tag{7.58}$$

and

$$\frac{D\overline{T}}{Dt} \approx \alpha\nabla\overline{T} + \nabla\cdot\Theta \tag{7.59}$$

where τ is the SPS stress and Θ is the SPS heat flux. They represent the contribution from subparticle scales to resolved scales for momentum transfer and heat transfer. Based on the filtering process, the definition of τ is as follows:

$$\tau_{\alpha\beta} = \overline{u}_\alpha \overline{u}_\beta - \overline{u_\alpha u_\beta} \qquad (7.60)$$

and the definition of Θ is as follows:

$$\Theta_\alpha = \overline{u}_\alpha \overline{T} - \overline{u_\alpha T} \qquad (7.61)$$

The SPS stress and heat flux must be modeled, because $\overline{u_\alpha u_\beta}$ or $\overline{u_\alpha T}$ is unknown. The most widely adopted Smagorinsky model for τ is as follows:

$$\tau_{\alpha\beta} = 2\nu_t \overline{S}_{\alpha\beta} - \frac{1}{3}\delta_{\alpha\beta}\tau_{\gamma\gamma} \qquad (7.62)$$

where the eddy viscosity ν_t is computed as follows:

$$\nu_t = (C_s\Delta)^2 \left(2\overline{S}_{\alpha\beta}\overline{S}_{\alpha\beta}\right)^{1/2} \qquad (7.63)$$

and the stain rate based on the filtered velocity $\overline{S}_{\alpha\beta}$ is calculated as follows:

$$\overline{S}_{\alpha\beta} = \frac{1}{2}\left(\frac{\partial \overline{u}_\alpha}{\partial x_\beta} + \frac{\partial \overline{u}_\beta}{\partial x_\alpha}\right) \qquad (7.64)$$

In Eqs. (7.62) to (7.64), $\delta_{\alpha\beta}$ is the Kronecker's delta, Δ is the cutoff scale, and C_s is the coefficient of the static model. The typical value of C_s is from 0.06 to 0.25 based on the authors' experience.

According to the derivation of Gotoh et al. [24], the SPS stress can also be written as

$$\tau_{\alpha\beta} = 2\nu_t \overline{S}_{\alpha\beta} - \frac{2}{3}\delta_{\alpha\beta}k_{SPS} \qquad (7.65)$$

where k_{SPS} is the SPS turbulence kinetic energy. This term can be approximated from the turbulence eddy viscosity as follows:

$$\nu_t = C_s k_{SPS}^{1/2}\Delta \qquad (7.66)$$

Then, the turbulence kinetic energy can be approximated by

$$k_{SPS} = \frac{C_s^4}{C_\nu^2}\Delta^2\left(2\overline{S}_{\alpha\beta}\overline{S}_{\alpha\beta}\right) \qquad (7.67)$$

where C_ν is an SPS turbulence constant. This provides a simpler manner to model the SPS stress term because no unknown is involved in the right-hand side of Eq. (7.62). It is noted that the unknown term $\frac{1}{3}\delta_{\alpha\beta}\tau_{\gamma\gamma}$ in Eq. (7.62) can be combined with the pressure calculation [24]. In this manner, this term can be directly omitted.

Similar to the SPS model, the SPS heat flux could be modeled as follows:

$$\Theta_\alpha = \alpha_T \frac{\partial \overline{T}}{\partial x} \tag{7.68}$$

and

$$\alpha_T = \frac{\nu_t}{\mathrm{Pr}_t} \tag{7.69}$$

where α_T is the SPS thermal diffusivity, and Pr_t is the SPS Prandtl number. The typical value of Pr_t is from 0.8 to 1.2 based on the authors' experience.

Regarding the discretization, only the gradient and divergence models are involved in the SPS models. Therefore, the presented discretization models in Chapters 2 and 3 can be directly adopted. It must be noted that there is no need to replace the center variables by a minimal value in the neighborhood (like the modified pressure gradient model for stability).

References

[1] Seiichi Koshizuka, Hirokazu Ikeda, Yoshiaki Oka, Numerical analysis of fragmentation mechanisms in vapor explosions, Nucl. Eng. Des. 189 (1999) 423−433.
[2] Hirokazu Ikeda, Seiichi Koshizuka, Yoshiaki Oka, Hyun Sun Park, Jun Sugimoto, Numerical analysis of jet injection behavior for fuel-coolant interaction using particle method, J. Nucl. Sci. Technol 38 (3) (2001) 174−182.
[3] Yuma Shimizu, Hitoshi Gotoh, Abbas Khayyer, An MPS-based particle method for simulation of multiphase flows characterized by high density ratios by incorporation of space potential particle concept, Comput. Math. Appl. 76 (2018) 1108−1129.
[4] Abbas Khayyer, Hitoshi Gotoh, Yuma Shimizu, A projection-based particle method with optimized particle shifting for multiphase flows with large density ratios and discontinuous density fields, Comput. Fluids 179 (2019) 356−371.
[5] Ahmad Shakibaeinia, Yee-Chung Jin, MPS mesh-free particle method for multiphase flows, Comput. Methods Appl. Mech. Engrg (2012) 13−26. 229−232.
[6] Abbas Khayyer, Hitoshi Gotoh, Enhancement of performance and stability of MPS mesh-free particle method for multiphase flows characterized by high density ratios, J. Comput. Phys 242 (2013) 211−233.
[7] Guangtao Duan, Bin Chen, Seiichi Koshizuka, Hao Xiang, Stable multiphase moving particle semi-implicit method for incompressible interfacial flow, Comput. Methods Appl. Mech. Eng. 318 (2017) 636−666.

[8] S. Hysing, S. Turek, D. Kuzmin, et al., Quantitative benchmark computations of two-dimensional bubble dynamics, Int. J. Num. Methods Fluids 60 (11) (2009) 1259−1288.

[9] Jianqiang Wang, Xiaobing Zhang, Improved Moving Particle Semi-implicit method for multiphase flow with discontinuity, Comput. Methods Appl. Mech. Eng. 346 (2019) 312−331.

[10] Mikio Sakai, Yusuke Shigeto, Xiaosong Sun, Takuya Aoki, Takumi Saito, Jinbiao Xiong, et al., Lagrangian−Lagrangian modeling for a solid−liquid flow in a cylindrical tank, Chem. Eng. J. (2012) 663−672. 200−202.

[11] M. Sakai, S. Koshizuka, Large-scale discrete element modeling in pneumatic conveying, Chem. Eng. Sci. 64 (2009) 533−539.

[12] M. Sakai, H. Takahashi, C.C. Pain, J. Latham, J. Xiang, Study on a large-scale discrete element model for fine particles in a fluidized bed, Adv. Powder Technol. 23 (2012) 673−681.

[13] Seiichi Koshizuka, Atsushi Nobe, Yoshiaki Oka, Numerical analysis of breaking waves using the moving particle semi-implicit method, Int. J. Numer. Methods Fluids 26 (1998) 751−769.

[14] E. Jafari Nodoushan, Ahmad Shakibaeinia, Khosrow Hosseini, A multiphase meshfree particle method for continuum-based modeling of dry and submerged granular flows, Powder Technol 335 (2018) 258−274.

[15] C.L. Chen, Comprehensive review of debris flow modelling concepts in Japan, Geol. Soc. Am. Rev. Eng. Geol. (1987) 13−29. VII.

[16] Mohanna Tajnesaie, Ahmad Shakibaeinia, Khosrow Hosseini, Meshfree particle numerical modelling of sub-aerial and submerged landslides, Comput. Fluids 172 (2018) 109−121.

[17] M. Jandaghian, A. Krimi, A. Shakibaeinia, Enhanced weakly-compressible MPS method for immersed granular flows, Adv. Water Resour. 152 (2021) 103908.

[18] E.Y.M. Lo, S. Shao, Simulation of near-shore solitary wave mechanics by an incompressible SPH method, Appl. Ocean Res. 24 (2002) 275−286.

[19] S. Shao, H. Gotoh, Turbulence particle models for tracking free surfaces, J. Hydraul. Res 43 (2005) 276−289.

[20] S. Shao, C. Ji, SPH computation of plunging waves using a 2-D sub-particle scale (SPS) turbulence model, Int. J. Numer. Methods Fluids. 51 (2006) 913−936.

[21] S. Shao, C. Ji, D.I. Graham, D.E. Reeve, P.W. James, A.J. Chadwick, Simulation of wave overtopping by an incompressible SPH model, Coast. Eng. 53 (2006) 723−735.

[22] J. Arai, S. Koshizuka, K. Murozono, Large eddy simulation and a simple wall model for turbulent flow calculation by a particle method, Int. J. Numer. Methods Fluids. 71 (2013) 772−787.

[23] G.T. Duan, B. Chen, Large Eddy Simulation by particle method coupled with Sub-Particle-Scale model and application to mixing layer flow, Appl. Math. Model. 39 (2015) 3135−3149.

[24] H. Gotoh, T. Shibahara, T. Sakai, Sub-particle-scale turbulence model for the MPS method - Lagrangian flow model for hydraulic engineering, Adv. Methods Comput. Fluid Dyn. (2001) 339−347. 9−4.

CHAPTER EIGHT

Heat transfer models

In this chapter the heat transfer and the associated boundary conditions including the Dirichlet, Neumann, and Robin boundaries are introduced; the advanced solid–liquid and gas–liquid phase change models are reviewed. In the modeling of solid–liquid phase change a new algorithm to deal with the numerical creep problem in simulation of highly viscous flow is presented. For the evaporation process the particle splitting procedure as the increase of gas volume is introduced and verified.

8.1 Governing equation and discretization

The governing equation for heat transfer is the energy conservation equation:

$$\rho \frac{DH}{Dt} = \nabla \cdot (k\nabla T) + Q \tag{8.1}$$

where H is the specific enthalpy, k is the thermal conductivity, T is the temperature, and Q is the heat source.

Omitting the heat source, Eq. (8.1) is solved explicitly, and using the classical moving particle semi-implicit (MPS) Laplacian model, Eq. (8.1) can be discretized by

$$\rho_i \frac{H_i^{n+1} - H_i^n}{\Delta t} = \frac{2d}{\lambda n^0} k_i \sum_j (T_j - T_i) w_{ij} \tag{8.2}$$

While treating with the conductive heat transfer between two different materials, harmonic mean of thermal conductivity is used and Eq. (8.2) is rewritten as

$$\rho_i \frac{H_i^{n+1} - H_i^n}{\Delta t} = \frac{2d}{\lambda n^0} \sum_j \frac{2k_i k_j}{k_i + k_j} (T_j - T_i) w_{ij} \tag{8.3}$$

Moving Particle Semi-implicit Method.
DOI: https://doi.org/10.1016/B978-0-443-13508-8.00008-1

High-order Laplacian model can also be used to discretize the conductive heat transfer term.

8.2 Heat transfer boundary conditions

The boundary condition is essential for heat transfer problems. In general the boundary conditions associated with the classical heat diffusion equations can be simply classified into three types: Dirichlet, Neumann, and Robin boundary conditions, which are also known as first type, second type, and third type boundary conditions, respectively. The temperature at the boundary is specified in a Dirichlet boundary condition, while the heat flux serves as a boundary condition in the Neumann/Robin cases. The Neumann boundary condition corresponds to a constant prescribed heat flux, while the heat flux is described as a linear/nonlinear function of boundary temperature in the Robin boundary condition.

In a Dirichlet boundary condition a fixed temperature is given at the boundary.

$$T_i = T_{\text{boundary}}, i \in \partial\Omega_{\text{boundary}} \tag{8.4}$$

The Neumann boundary condition can be written as

$$\frac{\partial T_i}{\partial \mathbf{n}} = C, i \in \partial\Omega_{\text{boundary}} \tag{8.5}$$

where C is a prescribed value; $\mathbf{n}$ is defined as the inward unit normal vector to the boundary $\partial\Omega_{\text{wall}}$.

When the heat convection is considered on the boundary the following equation should hold:

$$k\frac{\partial T_i}{\partial \mathbf{n}} = \gamma(T_i - T_\infty), i \in \partial\Omega_{\text{boundary}} \tag{8.6}$$

where γ represents the heat transfer coefficient on the boundary and T_∞ indicates the reference surrounding temperature. Finally the linear Robin boundary condition can be expressed as

$$\frac{\partial T_i}{\partial \mathbf{n}} = f(T_i) = AT_i + B, i \in \partial\Omega_{\text{boundary}} \tag{8.7}$$

where $f(T)$ is a linear function of temperature T to represent the heat flux and A and B are coefficients.

As mentioned in Chapter 5, for the particles near a boundary the support domain of the weight function (or kernel function) is cut off near the boundary, which makes it difficult to enforce accurate boundary conditions. Many methods have been proposed to enhance the accuracy. The so-called continuum surface reaction method was developed by Ryan et al. [1]. In their method the inhomogeneous Neumann/Robin boundary conditions are replaced by the homogeneous Neumann boundary. Meanwhile a volumetric source term is added to the governing equation. Another way is to extrapolate a polynomial fitted by the inner particles for the wall boundary particles, as shown in Fig. 8.1. For instance the Dirichlet boundary condition is given as

$$T_j = \left(1 + \frac{d_j}{d_i}\right) T_\Gamma - \frac{d_j}{d_i} T_i \qquad (8.8)$$

where T_Γ denotes the fixed value at the boundary Γ for the Dirichlet boundary condition. d_i and d_j indicate the normal distances from fluid particle i and wall particle j, respectively.

They are proven effective for various boundary conditions [1−3]. However, it could be troublesome to determine the specific positions of these ghost or boundary particles for complicated curved geometric shapes. Some preprocessing or boundary implementation might be a little complicated and inefficient, particularly for engineering problems. Moreover the above methods would suffer from some difficulty for the

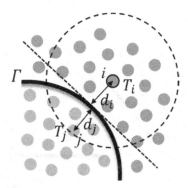

Figure 8.1 Extrapolating the temperature of inner particles for wall boundary particles near a solid boundary [2].

deformable and movable free-surface boundaries, considering the difficulty in arranging the ghost or boundary particles from time to time.

Recently the least squares method (e.g., LSMPS and MPS with a corrective matrix) has been introduced into the MPS method. The boundary particles can be directly defined on the boundary, making it much simpler to enforce the boundary condition. As mentioned in Chapter 5 the fixed boundary particle and the virtual boundary particle are the two widely adopted representations (see Fig. 8.2). The Dirichlet boundary condition can be treated simply by substituting computational variables on the boundary particle with prescribed boundary values.

In the fixed boundary model the Neumann and Robin boundaries can be directly imposed with boundary constraints (see Chapter 5) since the computational variable is defined on the boundary particle. Moreover the nonlinear Robin boundary (e.g., radiation boundary) can be solved using Newton's iteration. In this approach the specific positions of the boundary particles are needed to predefine. Careful attention should be paid to the complex geometry as well.

In the virtual boundary particle model, no computational variables are defined for the boundary particles, which are dynamically and locally generated on the boundaries (see Chapter 5). The treatments between Neumann and Robin boundary conditions would be pretty different. Assuming that the particle is located in the effective domain near the generated wall particle j the temperature of this wall particle, denoted by T_j, can be approximated using a Taylor series expansion:

$$T_j = T_i + \sum_{m=1}^{p} \frac{1}{m!} \left(\mathbf{r}_{ij} \cdot \nabla \right)^m T|_{\mathbf{r}=\mathbf{r}_i} + \mathrm{o}\left(\left| \mathbf{r}_{ij} \right|^{p+1} \right) \tag{8.9}$$

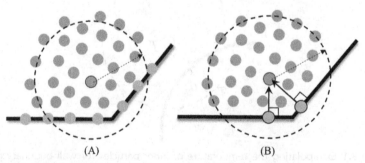

(A)   (B)

Figure 8.2 (A) Fixed and (B) virtual boundary particle representations.

If the higher-order terms are ignored, the following equation is obtained:

$$T_j - T_i = \sum_{m=1}^{p} \frac{r_s^m}{m!} \left(\frac{\mathbf{r}_{ij}}{r_s} \cdot \nabla \right)^m T|_{\mathbf{r}_i} = \mathbf{p}_{ij} \cdot \left(\mathbf{H}_{r_s}^{-1} \partial T|_{\mathbf{r}_i} \right) \tag{8.10}$$

where $\mathbf{r}_{ij} = \mathbf{r}_j - \mathbf{r}_i, r_s$ denotes a scaling parameter, p is a positive integer, which indicates the order of approximation, $\mathbf{p}_{ij}$ denotes a polynomial basis vector, $\mathbf{H}_{r_s}$ represents a scaling matrix, and ∂ indicates a differential operator vector.

For the second-order scheme ($p = 2$), these functions are written as

$$\mathbf{p}_{ij} = \mathbf{p}\left(\frac{\mathbf{r}_{ij}}{r_s} \right) = \left[\frac{x_{ij}}{r_s}, \frac{y_{ij}}{r_s}, \frac{x_{ij}^2}{r_s^2}, \frac{x_{ij}y_{ij}}{r_s^2}, \frac{y_{ij}^2}{r_s^2} \right]^T \tag{8.11}$$

$$\mathbf{H}_{r_s} = \mathrm{diag}\left(\frac{1}{r_s}, \frac{1}{r_s}, \frac{2}{r_s^2}, \frac{1}{r_s^2}, \frac{2}{r_s^2} \right) \tag{8.12}$$

$$\partial = \left[\frac{\partial}{\partial x}, \frac{\partial}{\partial y}, \frac{\partial^2}{\partial x^2}, \frac{\partial^2}{\partial x \partial y}, \frac{\partial^2}{\partial y^2} \right]^T \tag{8.13}$$

By applying the gradient operator on both sides the following equation is obtained:

$$\nabla T|_{\mathbf{r}=\mathbf{r}_j} = \nabla T|_{\mathbf{r}=\mathbf{r}_i} + \sum_{m=1}^{p-1} \frac{1}{m!} \left(\mathbf{r}_{ij} \cdot \nabla \right)^m \nabla T|_{\mathbf{r}=\mathbf{r}_i} \tag{8.14}$$

Multiplying both sides by the unit normal vector to the wall $\mathbf{n}_j$, Eq. (8.14) is transformed as

$$\frac{\partial T}{\partial \mathbf{n}}\bigg|_{\mathbf{r}=\mathbf{r}_j} = \mathbf{n}_j \sum_{m=0}^{p-1} \frac{r_s^m}{m!} \left(\frac{\mathbf{r}_{ij}}{r_s} \cdot \nabla \right)^m \nabla T|_{\mathbf{r}=\mathbf{r}_i} \tag{8.15}$$

On the other hand, substituting Eq. (8.10) into Eq. (8.7) gives

$$\frac{\partial T}{\partial \mathbf{n}}\bigg|_{\mathbf{r}=\mathbf{r}_j} = f(T_j) = AT_j + B = f(T_i) + A \sum_{m=1}^{p} \frac{r_s^m}{m!} \left(\frac{\mathbf{r}_{ij}}{r_s} \cdot \nabla \right)^m \nabla T|_{\mathbf{r}=\mathbf{r}_i} \tag{8.16}$$

Combining Eq. (8.16) and Eq. (8.15) the following equation can be obtained:

$$\mathbf{q}_{ij} \cdot \left(\mathbf{H}_{r_s}^{-1} \partial T|_{\mathbf{r}=\mathbf{r}_i} \right) \simeq r_s f(T_i) = r_s(AT_i + B) \tag{8.17}$$

where $\mathbf{q}_{ij} = \frac{\partial}{\partial \mathbf{n}}(r_s\mathbf{p}_{ij}) - Ar_s\mathbf{p}_{ij}$ indicates a modified polynomial basis vector. Specifically, $\mathbf{q}_{ij}$ for $p = 2$ is given as

$$\mathbf{q}_{ij} = \left[n_x - Ax_{ij}, n_y - Ay_{ij}, \frac{2n_x x_{ij} - Ax_{ij}^2}{r_s}, \frac{2n_y y_{ij} - Ay_{ij}^2}{r_s}, \frac{n_x y_{ij} + n_y x_{ij} - Ax_{ij} y_{ij}}{r_s}\right]^T$$

(8.18)

where (n_x, n_y) indicate the components of $\mathbf{n}_j$.

The objective function for the least squares problem is defined as

$$J(\mathbf{X}_i) = \sum_{j \neq i, j \in \Lambda_F} w_{ij}\left(\mathbf{p}_{ij} \cdot \mathbf{X}_i - T_j + T_i\right)^2 + \sum_{j \in \Lambda_B} w_{ij}\left(\mathbf{q}_{ij} \cdot \mathbf{X}_i - r_s f(T_i)\right)^2$$

(8.19)

with

$$\mathbf{X}_i = \mathbf{H}_{r_s}^{-1}\partial T|_{\mathbf{r}=\mathbf{r}_i}$$

(8.20)

where Λ_F and Λ_B indicate fluid particles and boundary particles, respectively. Minimization of J leads to the normal equations:

$$(\mathbf{M}_i + \mathbf{N}_i)\mathbf{X}_i = \mathbf{b}_i + \mathbf{c}_i$$

(8.21)

where

$$\mathbf{M}_i = \sum_{j \neq i, j \in \Lambda_F} w_{ij}\mathbf{p}_{ij} \otimes \mathbf{p}_{ij}$$

(8.22)

$$\mathbf{N}_i = \sum_{j \in \Lambda_B} w_{ij}\mathbf{q}_{ij} \otimes \mathbf{q}_{ij}$$

(8.23)

$$\mathbf{b}_i = \sum_{j \neq i, j \in \Lambda_F} w_{ij}\mathbf{p}_{ij}\left(T_j - T_i\right)$$

(8.24)

$$\mathbf{c}_i = r_s \sum_{j \in \Lambda_B} w_{ij}\mathbf{q}_{ij}(AT_i + B)$$

(8.25)

Therefore if $\mathbf{M}_i + \mathbf{N}_i$ is not singular, the spatial derivatives of temperature are obtained as

$$\partial T|_{\mathbf{r}=\mathbf{r}_i} = \mathbf{H}_{r_s}(\mathbf{M}_i + \mathbf{N}_i)^{-1}(\mathbf{b}_i + \mathbf{c}_i)$$

(8.26)

It is noted that the discrete equation does not include the temperature of the boundary particle $T_j(j \in \Lambda_B)$. The expected convergence rate is achieved with a simple test with complex geometry [4], as shown in Fig. 8.3. Fig. 8.4 presents an unsteady heat convection problem with an

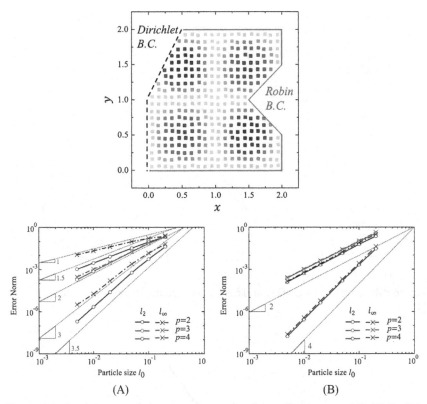

Figure 8.3 Convergence of the error norms for the verification test [4]. (A) Explicit test. (B) Implicit test.

internal heat source. As illustrated in Fig. 8.5 a reasonable convergence rate is obtained, even though the free-surface boundary moves over time due to surface tension.

8.3 Solid−liquid phase change model

The solid−liquid phase change includes the solidification and melting processes. These processes will greatly influence both the heat transfer and the fluid−solid interaction. The numerical models of phase change from the perspectives of heat transfer and fluid−solid interaction are discussed separately.

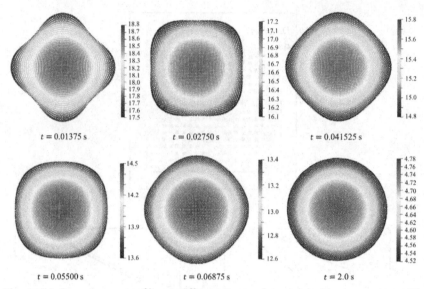

Figure 8.4 Temperature profiles at different time points, colored by temperature (K) [4].

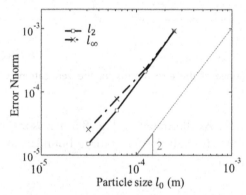

Figure 8.5 Global relative error l_2 and maximum relative error l_∞ of temperature at steady state [4].

For pure substances the phase change takes place at a constant temperature, namely, the melting temperature. During this process the enthalpy varies greatly, while the temperature remains at a constant one. For mixtures the phase change takes place in a range, namely, between the solidus and liquidus temperatures. When the phase change is numerically modeled, it is possible to consider that the phase change always occurs between the solidus and liquidus temperatures. In this situation the liquidus temperature is specified slightly higher than the solidus temperature for pure substances. The real values of solidus and liquidus temperatures of mixtures can be directly adopted.

There are two approaches to consider the phase-change effect on heat transfer. The first approach is to calculate the heat transfer explicitly based on the enthalpy so that the phase-change enthalpy can be directly calculated using Eq. (8.1). To consider the heat of fusion h_{sl} the relationship between the enthalpy h and the temperature T as follows:

$$h = \begin{cases} \rho C_{p,s} T & T < T_s \\ h_s + \dfrac{h_l - h_s}{T_l - T_s}(T - T_s) & T_s \leq T \leq T_l \\ h_l + \rho C_{p,l}(T - T_l) & T > T_l \end{cases} \tag{8.27}$$

where ρ is the density, $C_{p,s}$ is the specific heat capacity of the solid, $C_{p,l}$ is the specific heat capacity of the liquid, h_s is the enthalpy at the solidus temperature T_s, and h_l is the enthalpy at the liquidus temperatures T_l. The definition of h_s and h_l is as follows:

$$\begin{cases} h_s = \rho C_{p,s} T_s \\ h_l = h_s + \rho h_{sl} \end{cases} \tag{8.28}$$

It is usually assumed that the solid and liquid densities are the same in MPS simulation because the density variation (or volume variation) during the solid–liquid phase change is difficult to consider in MPS due to the manner to maintain incompressibility. Therefore a constant density ρ for both the solid and liquid phases is used above. After the enthalpy is updated explicitly based on Eq. (8.1) the temperature can be calculated as follows:

$$T = \begin{cases} \dfrac{h}{\rho C_{p,s}} & h < h_s \\ T_s + \dfrac{T_l - T_s}{h_l - h_s}(h - h_s) & h_s \leq h \leq h_l \\ T_l + \dfrac{(h - h_l)}{\rho C_{p,l}} & h > h_l \end{cases} \tag{8.29}$$

Then the solid fraction can be calculated from

$$
\gamma = \begin{cases} 1 & h < h_s \\ \dfrac{h_l - h}{h_l - h_s} & h_s \le h \le h_l \\ 0 & h > h_l \end{cases} \tag{8.30}
$$

In this manner the heat of fusion of phase change is incorporated into the heat transfer calculation. The main advantage is that the heat is exactly conserved during phase change if the original symmetric Laplacian model is employed. This approach is adopted in many studies [5−11].

The second approach is to convert the heat of fusion to a huge heat capacity during the phase change. For example the following heat transfer equation is used:

$$
\rho C_p \frac{dT}{dt} = k \nabla^2 T + q \tag{8.31}
$$

where C_p is not a constant but a variable dependent on the phase status. In this manner the enthalpy is not directly involved. To consider the heat of fusion h_{sl} the specific heat capacity is calculated based on temperature as follows:

$$
C_p = \begin{cases} C_{p,s} & T < T_s \\ \dfrac{h_{sl}}{T_l - T_s} & T_s \le T \le T_l \\ C_{p,l} & T > T_l \end{cases} \tag{8.32}
$$

After C_p is updated the temperature field can be solved explicitly or implicitly based on Eq. (8.31). Then the solid fraction can be calculated from

$$
\gamma = \begin{cases} 1 & T < T_s \\ \dfrac{T_l - T}{T_l - T_s} & T_s \le T \le T_l \\ 0 & T > T_l \end{cases} \tag{8.33}
$$

The main advantages of this approach are the simplicity and the capability to calculate heat transfer implicitly. When the fluid−solid interaction flow is secondary the implicit heat calculation can greatly save the computational cost [12]. However, it must be noted that this approach cannot exactly maintain the heat conservation. This approach was adopted in those studies of Refs. [13−17].

After the heat of fusion is incorporated into the heat-transfer calculation the approaches to consider the phase-change effect on fluid—solid
interaction are presented below. There are basically three approaches to
incorporate the phase-change effect, as presented below.

In the first approach the phase change is modeled by directly modifying the velocity of particles. When a fluid particle solidifies the particle is
immobilized in place by setting the velocity to zero directly. When a solid
particle is melted, this particle is allowed to move freely by changing the
solid particle to a fluid particle. To smooth the velocity change during the
phase change the particle velocity can be updated based on the solid fraction γ in the following manner:

$$\mathbf{u}' = \mathbf{u} \cdot \exp(- C \cdot \gamma) \tag{8.34}$$

where $\mathbf{u}'$ is the updated velocity, $\mathbf{u}$ is the original velocity, and C is a
dumping coefficient. The larger the coefficient C is the faster the
velocity decreases during solidification and vice versa. This approach
is adopted in many studies [6,18]. The main advantage of the approach
is the simplicity and high efficiency. It must be mentioned that this
approach is only suitable for the solidification near the wall, where the
solidified particles always attach to the wall. However, when the solidification takes place at free surfaces the solidified particles are still movable. In this situation the first approach cannot produce physical
results.

In the second approach the phase change is modeled by modifying the
viscosity of particles. Specifically, when a particle is subjected to solidification, its viscosity will be increased rapidly. On the other hand, when a
particle is subjected to melting, its viscosity will be decreased rapidly. The
commonly adopted model between the viscosity and solid fraction is the
Ramacciotti model [19]:

$$\mu(T) = \mu_{liq}\exp(2.5C\gamma(T)) \tag{8.35}$$

where μ_{liq} is the viscosity of the liquid phase and C is a key model coefficient
($C = 3.5 - 8$) [20] for solidification. Because the viscosity is rather high, the
viscosity term must be calculated implicitly in this situation [21,22]. The main
problem of this approach is the so-called "numerical creeping" issue. For
example, even though the viscosity of the solidified phases is increased to a
rather huge value, the solidified phases can still move gradually (i.e., numerical
creeping). Particularly the creeping velocity does not decrease continuously

with the increase of viscosity [22]. This is an unphysical phenomenon for the highly viscous flow.

Duan et al. [22] carefully investigated this phenomenon and found that the delayed diffusion calculation of the velocity change caused by the pressure gradient term is the reason for the numerical creeping. As shown in Fig. 8.6A the pressure terms are calculated after the viscosity terms in the original MPS algorithm (or most time integral algorithm for particle or mesh methods). In this situation the velocity change due to the pressure gradient term can still result in some creeping flow, even though the viscosity is quite large. In order to calculate the viscosity terms precisely, it is necessary to add another diffusion calculation for the velocity change due to the pressure gradient, as shown in Fig. 8.6B. Inspired by this observation, Duan et al. [22] proposed a new time integral algorithm for the highly viscous flow, as shown in Fig. 8.6C. In the new algorithm the viscosity diffusion is calculated last, guaranteeing that all the velocities and velocity changes are diffused by viscosity promptly. In this manner the numerical-creeping issue is avoided, and the viscosity diffusion can be calculated precisely. The comparison of the qualitative and quantitative results between the original and new algorithms is presented in Fig. 8.7. The simulated phenomenon is the melt spreading terminated by the crust formation (i.e., the solidification at free surfaces). As shown in Fig. 8.7A and B the spreading length simulated by the new algorithm is shorter than

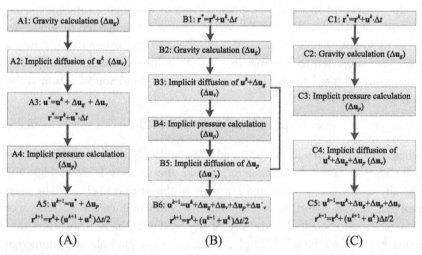

Figure 8.6 Sketch showing the comparison between the original and new algorithms [22]. (A) original algorithm, (B) temporary algorithm, and (C) new algorithm.

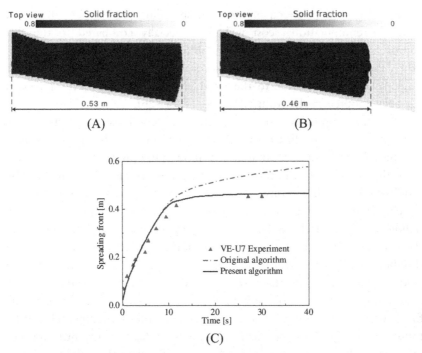

Figure 8.7 Result comparison between the original and new algorithms in the simulations of melt spreading [22]. (A) Debris distribution using the original algorithm. (B) Debris distribution using the new algorithm. (C) Histories of the spreading fronts.

that by the original algorithm. Meanwhile the result simulated by the new algorithm agrees better with the experimental measurements, as shown in Fig. 8.7C. More importantly, after 10 seconds the solidified melt front suffers from numerical creeping when the original algorithm is adopted. On the other hand, when the new algorithm is employed the high viscosity of the solidified crust can effectively terminate the spreading. In brief the new algorithm plays a key role in modeling the phase change via modifying the viscosity. When the new algorithm is adopted the modification of viscosity can effectively simulate the influence of solid—liquid phase change on the fluid—solid interaction flows.

The main feature of the second approach is summarized below. This approach can simulate the movable solidified phases at free surfaces or inside the melt. Meanwhile the strain rate of the solidified phases can be reliably obtained because the viscosity terms are precisely calculated. Based on the strain rate a crust fracture model can be developed to simulate the "stop-and-go" phenomena in the melt spreading [13,16]. On the other hand the computational cost is high, especially compared to the first approach. In addition, when two or multiple solidified phases touch each other, they will always automatically combine each other to form a large single solidified phase due to the rapid viscosity diffusion. This approach has been adopted in the studies [12,14−17,23−26].

The third approach can be regarded as a combination of the above two approaches. The first approach models the phase change by modifying the velocity. The second approach models the phase change by modifying the viscosity. In the third approach the phase change is modeled by modifying both the viscosity and velocity. Specifically, melting is modeled by changing a solid particle to a fluid one and reducing the viscosity. Solidification is modeled by increasing the viscosity to a huge value. Particularly, when the velocity of a solidified particle is smaller than a threshold, this particle will be immobilized [12,17]. The reason is that the velocity will quickly decrease to a tiny value when the solidified phase is near a wall. Otherwise the solidified phase is still movable if it is at the free surfaces or inside the melt. More details of the third approach can be found in [12,17]. This approach takes the advantages of the first and second approaches above. For example, it can simulate the solidification at free surfaces compared to the first approach and greatly save the computational cost compared to the second approach. However, this approach may misdetect the immobilized solidified phase sometimes, resulting in the solidified phase in the air [12].

Last, some key challenges of the solid−liquid phase-change model are pointed out after the basic models are presented. The aforementioned phase-change models are expected work reasonably only in some relatively simple flows. In complicated flows with phase changes (1) the breakup of solidified phases due to the collision or violent fluid−solid interactions and (2) the merging of solidified phases due to the newly solidified particles can take place. To model the breakup of solidified phases the mechanical stress analysis of the solidified phases must be considered. To model the merging of the solid phases, one must distinguish the already solidified particles and the newly- solidified particles, considering that the high viscosity model for the solidified phases can always overestimate the merging phenomena. To our best knowledge the current development of numerical approaches is still far away to reasonably model the complicated breakup and merging phenomena of the solidified phases.

8.4 Gas–liquid phase change model

Different from solid–liquid phase change, volume change has to be considered in gas–liquid phase change model. The earliest study was proposed by Yoon et al. [27] to calculate bubble growth in nucleate pool boiling. In their work a uniform temperature and pressure are assumed for the vapor phase so that the field equations were solved only for the liquid phase. The same model was used to investigate bubble dynamics during flow boiling [28]. In their model, gas phase was not calculated directly. Liu et al. [29] developed a gas–liquid phase change model in which both the gas and liquid phases are solved directly.

In this vaporization model, each particle is associated with a set of variables, namely, temperature, enthalpy, and liquid volume fraction. Physical properties such as density and thermal conductivity vary with vapor volume fraction as

$$\rho_i = \alpha_i \rho_{\text{vapor}} + (1 - \alpha_i)\rho_{\text{liquid}} \qquad (8.36)$$

$$k_i = \alpha_i k_{\text{vapor}} + (1 - \alpha_i)k_{\text{liquid}} \qquad (8.37)$$

$$\mu_i = \alpha_i \mu_{\text{vapor}} + (1 - \alpha_i)\mu_{\text{liquid}} \qquad (8.38)$$

Here α_i is the vapor volume fraction of particle i.

If particle enthalpy exceeds the liquidus enthalpy, mass transfer due to gas–liquid phase change is calculated as

$$\Delta m_i^n = \rho_i V_i \frac{H_i^{n+1} - H_i^n}{H_{\text{latent}}} \qquad (8.39)$$

where Δm_i is the total amount of mass transfer due to phase change and H_{latent} is the latent heat of vaporization. Volume change due to phase change of particle i is

$$\Delta V_i = \Delta m_i^n \left(\frac{1}{\rho_{\text{vapor}}} - \frac{1}{\rho_{\text{liquid}}} \right) \qquad (8.40)$$

Next the mass conservation equation is reformulated to count in the mass transfer due to phase change. If there is no phase change in incompressible flow, the density is constant and the mass conservation equation is equal to the following velocity divergence-free condition:

$$\nabla \cdot \mathbf{u} = 0 \qquad (8.41)$$

However, during the vaporization process the volume varies due to the density difference of liquid and vapor and the divergence-free condition for the velocity field is not satisfied. As the divergence of the velocity field equals the time rate of change of the particle volume, one can obtain

$$\langle \nabla \cdot \mathbf{u} \rangle_i^{n+1} = \frac{\Delta V_i^n}{V_i^n \Delta t} = \frac{\Delta m_i^n}{V_i^n \Delta t} \left(\frac{1}{\rho_{\text{vapor}}} - \frac{1}{\rho_{\text{liquid}}} \right) \tag{8.42}$$

Accordingly a new pressure Poisson equation (PPE) with the source term of mass transfer gives

$$\nabla \cdot \left(\frac{1}{\rho} \nabla p^{n+1} \right) = \frac{1}{\Delta t} \left[\nabla \cdot \mathbf{u}^* - \frac{\Delta m_i^n}{V_i^n \Delta t} \left(\frac{1}{\rho_{\text{vapor}}} - \frac{1}{\rho_{\text{liquid}}} \right) \right]. \tag{8.43}$$

During the vaporization process the particle volume increases according to Eq. (8.40). An increased volume results in an increased distance between particles near the interface. The particle splitting technique is necessary to avoid the instability caused by large volume difference.

The splitting procedure is described in Fig. 8.8. If the volume of particle $iV_i > \eta V_{ini}$, particle i is split to two children particles, child particle 1 and child particle 2, along the direction of the interface norm vector at particle i. Here η is set to 1.2 and V_{ini} is the initial particle volume. In two-dimensional calculation, $V_{ini} = l_0^2$. The two children particles are two particles with equal volumes, half of the parent particle volume. Vapor volume fraction change is considered in the splitting procedure and is illustrated in Fig. 8.9. In Fig. 8.9 the shaded portion of the particles is occupied by vapor. If the vapor volume fraction of parent particle i is smaller than 0.5, one of the children particles is a liquid particle and the vapor contained in parent particle i is allocated to the other child particle. If the vapor volume fraction of parent particle i is larger than 0.5, one of the children particles is the vapor particle and the other particle contains a

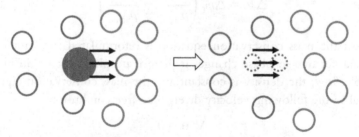

Figure 8.8 Description of the splitting procedure in the vaporizaiton model.

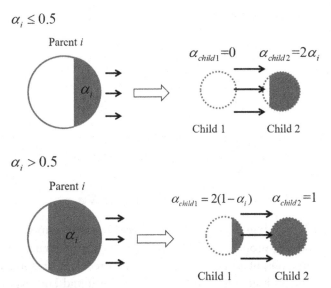

Figure 8.9 Vapor volume fraction change in the splitting procedure in the vaporizaiton model [29].

liquid—vapor mixture. The vapor volume fraction of the two children particles is calculated correspondingly:

$$\begin{aligned} \alpha_i \leq 0.5 : \quad & \alpha_{\text{child1}} = 0 \qquad \alpha_{\text{child2}} = 2\alpha_i \\ \alpha_i > 0.5 : \quad & \alpha_{\text{child1}} = 2(1 - \alpha_i) \quad \alpha_{\text{child2}} = 1 \end{aligned} \tag{8.44}$$

Physical properties such as density, viscosity coefficient, and thermal conductivity can be calculated from the vapor volume fraction:

$$\rho_{\text{child}} = \alpha_{\text{child}}\rho_{\text{vapor}} + (1 - \alpha_{\text{child}})\rho_{\text{liquid}} \tag{8.45}$$

$$k_{\text{child}} = \alpha_{\text{child}}k_{\text{vapor}} + (1 - \alpha_{\text{child}})k_{\text{liquid}} \tag{8.46}$$

$$\mu_{\text{child}} = \alpha_{\text{child}}\mu_{\text{vapor}} + (1 - \alpha_{\text{child}})\mu_{\text{liquid}} \tag{8.47}$$

The specific enthalpies of the two children particles are calculated by

$$\alpha_i \leq 0.5 : H_{\text{child1}} = H_{\text{liquid}}, H_{\text{child2}} = \frac{V_{\text{child2,liquid}}\rho_{\text{liquid}}H_{\text{liquid}} + V_{\text{child2,vapor}}\rho_{\text{vapor}}H_{\text{vapor}}}{V_{\text{child2}}\rho_{\text{child1}}}$$
$$\tag{8.48}$$

$$\alpha_i > 0.5 : H_{\text{child2}} = H_{\text{vapor}}, H_{\text{child1}} = \frac{V_{\text{child1,liquid}}\rho_{\text{liquid}}H_{\text{liquid}} + V_{\text{child1,vapor}}\rho_{\text{vapor}}H_{\text{vapor}}}{V_{\text{child1}}\rho_{\text{child1}}}$$
$$\tag{8.49}$$

Position vectors of the two children particles are

$$\mathbf{r}_{\text{child1}} = \mathbf{r}_{\text{parent}} + \frac{1}{2}\sqrt{\frac{V_i}{2}}\mathbf{n}_i \tag{8.50}$$

$$\mathbf{r}_{\text{child2}} = \mathbf{r}_{\text{parent}} - \frac{1}{2}\sqrt{\frac{V_i}{2}}\mathbf{n}_i \tag{8.51}$$

Here $\mathbf{n}_i$ is the unit normal vector along the direction of temperature gradient. The velocity of the child particle is obtained via a second-order interpolation:

$$\varphi_{\text{child}} = \varphi_{\text{parent}} + (\mathbf{r}_{\text{child}} - \mathbf{r}_{\text{parent}}) \cdot \nabla\varphi_{\text{parent}} \tag{8.52}$$

Using this gas—liquid phase change model the two-dimensional horizontal film boiling problem is studied. The schematic diagram is shown in Fig. 8.10. The computations are performed in a rectangular domain that is periodic in the x-direction. The densities of the liquid and vapor are set to 200.0 and 5.0 kg/m³, respectively. The specific heats of the liquid and vapor are set to 400 and 200 J/(kg·K), respectively. The thermal conductivities of the liquid and vapor are set to 40.0 and 1.0 W/(m·K), respectively. The surface tension of 0.1 N/m and the latent heat of vaporization of 10 kJ/kg are used in this liquid and vapor system. The temperature profile is initialized as a simple linear profile between the liquid and wall

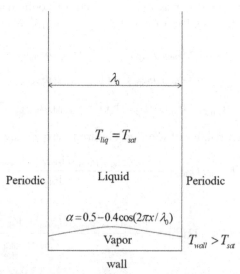

Figure 8.10 Schematic diagram of two-dimensional horizontal film boiling [29].

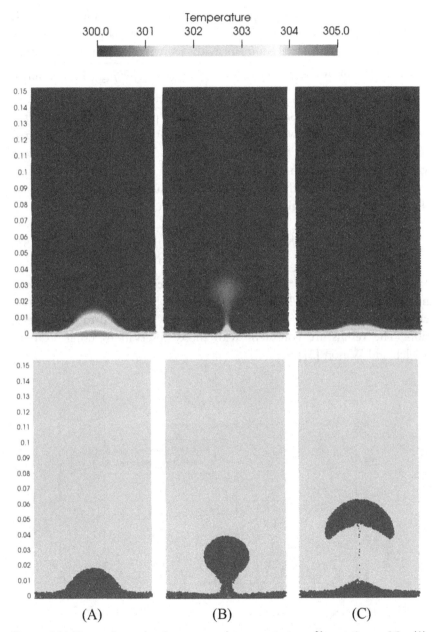

Figure 8.11 Vapor shape development and temperature profiles at time = 0.3 s (A), 0.4 s (B), and 0.5 s (C) in the simulation of two-dimensional film boiling using 100 × 300 particles [29].

temperatures. The liquid is kept under a saturation condition with 300K, while the wall temperature is 5K higher than the saturation temperature. We use a computational domain with its width equal to the most unstable Taylor wavelength: $\lambda_0 = 2\pi\sqrt{3\sigma/g(\rho_{\text{liquid}} - \rho_{\text{vapor}})}$. The initial two-phase interface is initially located at $y = (4 + \cos(2\pi x/\lambda_0))\lambda_0/100$.

Vapor shape development and temperature profiles at time = 0.3, 0.4, and 0.5 second are shown in Fig. 8.11. The saturated liquid is vaporized due to the heat transfer from the heated vapor. Vapor film grows up to a vapor bubble and detaches from the bottom.

To check the heat flux the local Nusselt number is calculated as the dimensionless heat flux through the wall:

$$Nu = \frac{\lambda_c}{(T_{\text{wall}} - T_{\text{sat}})}\frac{\partial T}{\partial y}\bigg|_{y=0} \tag{8.53}$$

Here $\lambda_c = \sqrt{\sigma/g(\rho_{\text{liquid}} - \rho_{\text{vapor}})}$ is the characteristic length. The Nusselt number is compared to Klimenko's correlation [30] in Fig. 8.12. The calculated Nu shows some fluctuating discrepancy compared to the correlation. The results are consistent with those observations using VOF in Welch and Wilson [31].

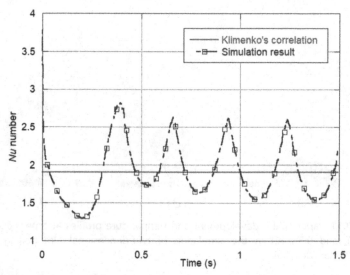

Figure 8.12 Nussel number in the simulation of two-dimensional film boiling [29].

References

[1] E.M. Ryan, A.M. Tartakovsky, C. Amon, A novel method for modeling Neumann and Robin boundary conditions in smoothed particle hydrodynamics, Comp. Phys. Commun. 181 (12) (2010) 2008−2023.

[2] J. Wang, W. Hu, X. Zhang, W. Pan, Modeling heat transfer subject to inhomogeneous Neumann boundary conditions by smoothed particle hydrodynamics and peridynamics, Int. J. Heat. Mass. Transf. 139 (2019) 948−962.

[3] W. Pan, J. Bao, A.M. Tartakovsky, Smoothed particle hydrodynamics continuous boundary force method for Navier−Stokes equations subject to a Robin boundary condition, J. Comput. Phys. 259 (2014) 242−259.

[4] Z. Wang, G. Duan, T. Matsunaga, T. Sugiyama, Consistent Robin boundary enforcement of particle method for heat transfer problem with arbitrary geometry, Int. J. Heat. Mass. Transf. 157 (2020) 119919.

[5] T. Kawahara, Y. Oka, Ex-vessel molten core solidification behavior by moving particle semi-implicit method, J. Nucl. Sci. Technol. 49 (2012) 1156−1164.

[6] Y. Yasumura, A. Yamaji, M. Furuya, Y. Ohishi, G. Duan, Investigation on influence of crust formation on VULCANO VE-U7 corium spreading with MPS method, Ann. Nucl. Energy. 107 (2017) 119−127.

[7] R. Chen, L. Chen, K. Guo, A. Yamaji, M. Furuya, W. Tian, et al., Numerical analysis of the melt behavior in a fuel support piece of the BWR by MPS, Ann. Nucl. Energy. 102 (2017) 422−439.

[8] X. Li, Y. Oka, Numerical simulation of the SURC-2 and SURC-4 MCCI experiments by MPS method, Ann. Nucl. Energy. 73 (2014) 46−52.

[9] X. Li, A. Yamaji, Three-dimensional numerical study on the mechanism of anisotropic MCCI by improved MPS method, Nucl. Eng. Des. 314 (2017) 207−216.

[10] R. Chen, Y. Oka, G. Li, T. Matsuura, Numerical investigation on melt freezing behavior in a tube by MPS method, Nucl. Eng. Des. 273 (2014) 440−448.

[11] G. Li, Y. Oka, M. Furuya, Experimental and numerical study of stratification and solidification/melting behaviors, Nucl. Eng. Des. 272 (2014) 109−117.

[12] G. Duan, A. Yamaji, M. Sakai, A multiphase MPS method coupling fluid−solid interaction/phase-change models with application to debris remelting in reactor lower plenum, Ann. Nucl. Energy. 166 (2022) 108697.

[13] G. Duan, A. Yamaji, S. Koshizuka, A. Novel, Approach for Crust Behaviors in Corium Spreading Based on Multiphase MPS Method, in: 12th International Topical Meeting on Nuclear Reactor Thermal-Hydraulics, Operations and Safety, China, 2018, pp. 1−13.

[14] Jubaidah, G. Duan, A. Yamaji, C. Journeau, L. Buffe, J.F. Haquet, Investigation on corium spreading over ceramic and concrete substrates in VULCANO VE-U7 experiment with moving particle semi-implicit method, Ann. Nucl. Energy. 141 (2020) 107266.

[15] N. Takahashi, G. Duan, A. Yamaji, X. Li, I. Sato, Development of MPS method and analytical approach for investigating RPV debris bed and lower head interaction in 1F Units-2 and 3, Nucl. Eng. Des. 379 (2021) 111244.

[16] Jubaidah, Y. Umazume, N. Takahashi, X. Li, G. Duan, A. Yamaji, 2D MPS method analysis of ECOKATS-V1 spreading with crust fracture model, Nucl. Eng. Des. 379 (2021) 111251.

[17] R. Kawakami, X. Li, G. Duan, A. Yamaji, I. Sato, T. Suzuki, Improvement of solidification model and analysis of 3D channel blockage with MPS method, Front. Energy. (2021) 1−13.

[18] G. Li, M. Liu, G. Duan, D. Chong, J. Yan, Numerical investigation of erosion and heat transfer characteristics of molten jet impinging onto solid plate with MPS-LES method, Int. J. Heat. Mass. Transf. 99 (2016) 44−52.
[19] M. Ramacciotti, C. Journeau, F. Sudreau, G. Cognet, Viscosity models for corium melts, Nucl. Eng. Des. 204 (2001) 377−389.
[20] C. Journeau, E. Boccaccio, C. Brayer, G. Cognet, J.F. Haquet, C. Jégou, et al., Ex-vessel corium spreading: results from the VULCANO spreading tests, Nucl. Eng. Des. 223 (2003) 75−102.
[21] X. Sun, M. Sakai, K. Shibata, Y. Tochigi, H. Fujiwara, Numerical modeling on the discharged fluid flow from a glass melter by a Lagrangian approach, Nucl. Eng. Des. 248 (2012) 14−21.
[22] G. Duan, A. Yamaji, S. Koshizuka, A novel multiphase MPS algorithm for modeling crust formation by highly viscous fluid for simulating corium spreading, Nucl. Eng. Des. 343 (2019) 218−231.
[23] N. Takahashi, G. Duan, M. Furuya, A. Yamaji, Analysis of hemispherical vessel ablation failure involving natural convection by MPS method with corrective matrix, Int. J. Adv. Nucl. React. Des. Technol. 1 (2019) 19−29.
[24] X. Li, A. Yamaji, A numerical study of isotropic and anisotropic ablation in MCCI by MPS method, Prog. Nucl. Energy. 90 (2016) 46−57.
[25] R. Chen, Q. Cai, P. Zhang, Y. Li, K. Guo, W. Tian, et al., Three-dimensional numerical simulation of the HECLA-4 transient MCCI experiment by improved MPS method, Nucl. Eng. Des. 347 (2019) 95−107.
[26] G. Li, P. Wen, H. Feng, J. Zhang, J. Yan, Study on melt stratification and migration in debris bed using the moving particle semi-implicit method, Nucl. Eng. Des. 360 (2020) 110459.
[27] H.Y. Yoon, S. Koshizuka, Y. Oka, Direct calculation of bubble growth, departure and rise in nucleate pool boiling, Int. J. Multiph. Flow. 27 (2001) 277−298.
[28] R.H. Chen, W.X. Tian, G.H. Su, S.Z. Qiu, Y. Ishiwatari, Y. Oka, Numerical investigation on bubble dynamics during flow boiling using moving particle semi-implicit method, Nucl. Eng. Des. 240 (2010) 3830−3840.
[29] X. Liu, K. Morita, S. Zhang, Direct numerical simulation of incompressible multiphase flow with vaporization using moving particle semi-implicit method, J. Comput. Phys. 425, 109911.
[30] V.V. Klimenko, Film boiling on a horizontal plate-new correlation, Int. J. Heat. Mass. Transf. 24 (1981) 69−79.
[31] S.W.J. Welch, J.A. Wilson, Volume of fluid based method for fluid flows with phase change, J. Comput. Phys. 160 (2) (2000) 662−682.

CHAPTER NINE

Efficiency improvement

Particle methods are usually more time-consuming than mesh methods because of the larger interaction radius and frequently updated neighbor connections. Thus improving the calculation efficiency is a key issue in particle methods. Four kinds of acceleration methods, including OpenMP (open multiprocessing) parallelization, MPI (message passing interface) parallelization, GPU (graphics processing unit) parallelization, and multi-resolution models, are presented in this chapter. The simple OpenMP and complicated MPI are mature techniques to accelerate the moving particle semi-implicit (MPS) method. The GPU parallelization and multiresolution models under development are believed to greatly improve the computational efficiency.

9.1 OpenMP parallelization

The modern central processing units (CPUs) are composed of multiple cores. To fully exploit the performance of a multicore CPU, a multi-core parallelization is indispensable. The OpenMP model is designed for this purpose. It is very simple and yet effective to accelerate the code on a multicore workstation. In this chapter the basic ideas of OpenMP and simple program examples will be discussed below.

The OpenMP model is suitable for a workstation, where the multiple CPU cores can access the same shared memory, as shown in Fig. 9.1.

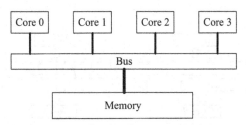

Figure 9.1 The hardware structure for the OpenMP model [1].

Moving Particle Semi-implicit Method.
DOI: https://doi.org/10.1016/B978-0-443-13508-8.00009-3

The shared memory can be used to exchange information among the different cores. The basic parallelization concept of OpenMP is shown in Fig. 9.2. Specifically the whole program is divided into a series of the series and parallel regions. In the series region, only one process (or the master thread) is running, and only one CPU core is used. In this region the program is executed like the normal series code. When the program enters the parallel region the process will fork several threads to execute the code block concurrently and simultaneously. In this manner the computational cost of the parallel region can be greatly saved. After the parallel region is already finished, all the threads will join into a single process. Based on the "fork-join" approach the program can be parallelized partially. When more code blocks are executed parallelly the parallel efficiency becomes higher. Practically, it is possible to only parallelize the bottleneck part of the program.

To explain how the parallel region is executed, an example of task assignment in parallelization is shown in Fig. 9.3. In particle methods, all the particles usually execute the same operations. For example, we have 16 particles and 4 CPU cores to perform the simulation. The OpenMP model will assign the particles to the cores based on the particle labels. As shown in Fig. 9.3, one possible task-assignment way is to assign the first four particles to the first core, the second four particles to the second

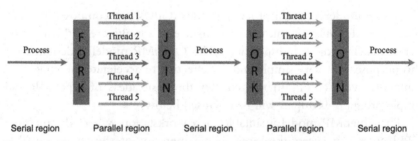

Figure 9.2 The parallelization conception of the OpenMP model.

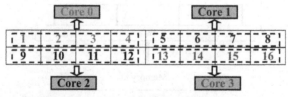

Figure 9.3 The task assignment in the OpenMP model.

core, and so on. In this manner the particles are assigned to the cores in a simple yet roughly load-balanced way.

After the basic concept of OpenMP is described an example in C language is presented below. The code computes the summation of the squares of the element of an array. The code is as follows:

```c
/*example_omp1.c*/
#include <stdio.h>
#include <omp.h>
int main()
{
    int i, a[15]={1,2,3,4,5,6,7,8,1,2,3,4,5,6,7}, sum=0;
    omp_set_num_threads(4);
#pragma omp parallel for schedule(static,4) private(i) shared(a) reduction(+:sum)
    for(i=0;i<15;i++)
    {
        a[i]=a[i]*a[i];
        printf("a[%d]=%d from core:%d\n", i, a[i], omp_get_thread_num());
        sum+=a[i];
    }
    printf("summation is %d\n",sum);
    return(0);
}
```

The "omp.h" is the header file for the OpenMP parallelization. The function "omp_set_num_threads()" is used to set how many threads (or cores) are utilized in the calculation. The directive "#pragma omp parallel for schedule(static,4) private(i) shared(a) reduction(+:sum)" is the key to determine how the OpenMP parallelization is performed. Specifically the clause "schedule" determines how to assign the tasks: the parameter "static" indicates that all the threads are allocated iterations before they execute the loop iterations; chunk size "4" means that the program will allocate four continuous elements of array "a" to the threads one by one. The clause "private" indicates which variables are private to a thread. The clause "shared" indicates which variables are shared among the threads. The clause "reduction" provides a safe way of joining work from all threads after the parallel region. Among these clauses, "private" and "shared" can be directly neglected in most cases.

When the code is compiled the openmp option must be enabled. The output of the above program is as follows:

a[0]=1 from core:0
a[1]=4 from core:0
a[2]=9 from core:0
a[12]=25 from core:3
a[13]=36 from core:3
a[14]=49 from core:3
a[3]=16 from core:0
a[8]=1 from core:2
a[9]=4 from core:2
a[10]=9 from core:2
a[11]=16 from core:2
a[4]=25 from core:1
a[5]=36 from core:1
a[6]=49 from core:1
a[7]=64 from core:1
summation is 344

It must be noted that the output of the elements is random because the "printf" clause is performed parallelly. Nevertheless, the final summation result is always correctly calculated.

After this simple example, we show how the particle number density of each particle can be calculated concurrently using the OpenMP model. The pseudocode is as follows:

```
/*example_omp2.c*/

void calculate_particle_number_density(int numberOfParticles, double *pnd)

{

#pragma omp parallel for schedule(dynamic, 50)

    for(int ip=0; ip<numberOfParticles; ip++)

    {

        double sum=0.0;

        /*loop the neighbor list*/

        for(int jp=0; jp<numberOfParticles; jp++)

        {

            /*r_ij is the distance between particle i and j; re is the interaction radius*/

            If(r_ij<re) sum+=weight(r_ij, re);

        }

        pnd[ip]=sum;

    }

}
```

In the above code the "dynamic" schedule is used because the variable number of particles may change during the simulations.

Briefly the advantages of the OpenMP model are the simplicity and efficiency. For example, one just made a minor modification to the original code and can greatly accelerate the program. However, this model can only be applied to one workstation. For example the speedup ratio is limited by the core number of the workstation. The purchase of a workstation with many cores is usually expensive and even unpractical.

9.2 Message passing interface parallelization

The MPI model is designed for the large-scale parallelization using supercomputers. The basic conception of the MPI model is shown in Fig. 9.4. Specifically, many processes work together to accomplish one job. In other words the job must be carefully assigned to various processes. Because each process has its own memory space, the communication (i.e., information exchange) among the processes becomes the key in the MPI parallelization. MPI actually defines a lot of interfaces/functions to transfer/exchange the information/messages among different processes. It is noted that these processes can be executed on the same workstation or particularly on different workstations. In this regard, MPI is especially suitable for clusters.

To further explain the concept of MPI a simple "hello world" example using the C language is shown below.

```c
/*example_mpi1.c*/

#include<stdio.h>

#include<mpi.h>

int main(int argc, char *argv[])

{

  int totalProcessNum, rankID;

  int rt = MPI_Init(&argc, &argv);

  MPI_Comm_size(MPI_COMM_WORLD, &totalProcessNum);

  MPI_Comm_rank(MPI_COMM_WORLD, &rankID);

  printf("Hello,    world!    %dth   of    totalProcessNum    =    %d\n",   rankID,

totalProcessNum);

  MPI_Finalize();

  return 0;

}
```

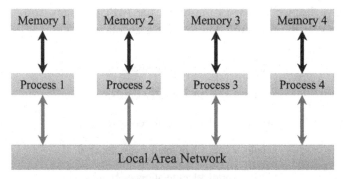

Figure 9.4 The parallelization conception of the MPI model.

The standard file "mpi.h" is the header file of MPI, which must be included when the MPI parallelization is employed. All the MPI functions have a prefix "MPI." The function "MPI_Init()," will initialize the MPI environment for the current process. Correspondingly, "MPI_Finalize()" will make the program to exit the MPI environment. All the processes must call MPI_Init() at the beginning of the program and call MPI_Finalize() at the end of the program. The parameter "MPI_COMM_WORLD," is the default communication environment (i.e., communicator), from which all the processes can communicate with each other. In the communication environment, every process has a unique ID, which is used to distinguish a specific process from other processes. The function "MPI_Comm_size()," is used to get the number of all the processes. The function "MPI_Comm_rank()," is used to get the ID or rank of a process. The "printf" clause will output the ID of the current process and the total number of processes.

To compile the above code an MPI compiler must be used. On a Linux or Unix system the following command can be used to compile the above program:

mpicc -o a.out example_mpi1.c

To start the program a.out using six processes the following command can be used:

mpirun −n 6 ./a.out

Briefly, "mpicc" is used to compile the MPI program, and "mpirun" is used to start the program. The output is as follows:

```
Hello, world! 0th of totalProcessNum = 6

Hello, world! 2th of totalProcessNum = 6

Hello, world! 5th of totalProcessNum = 6

Hello, world! 1th of totalProcessNum = 6

Hello, world! 4th of totalProcessNum = 6

Hello, world! 3th of totalProcessNum = 6
```

In the output the process ID is not in a serial order because the six processes print the message simultaneously. Thus the output is in a random manner. The MPI program is executed in the following manner. When the command "*mpirun −n 6 ./a.out*" is submitted, six processes will execute the program a out concurrently. Before the function MPI_Init() and after the function MPI_Finalize(), all the processes execute exactly the same commands. After the function MPI_Init(), each process will have a unique ID and still execute the same codes. Nevertheless, different processes will perform different tasks based on the process ID. It is noted that the different tasks for different processes are usually written in the same MPI program. Therefore the key of an MPI program is to (1) divide a whole job into different tasks and (2) assign the different tasks to different processes based on the process ID.

When the MPI model is used to parallelize a simulation of the particle method the domain decomposition method [2,3] is usually adopted. As shown in Fig. 9.5, first the whole domain is divided into several subdomains. Second the subdomains are one by one assigned to the processes. Every process only calculates the motion of particles which are located in the assigned subdomain.

After the domain decomposition, communication and load balance become the main issues in parallelizing the MPS method by the MPI model. As shown in Fig. 9.5 a background grid whose size is exactly the same as the interaction radius of a particle is established for the linked-list algorithm to search neighbor particles. Therefore a reference particle only interacts with the particles which are located in the nine cells close to the reference particle in 2D. When the motion of a particle which is close to

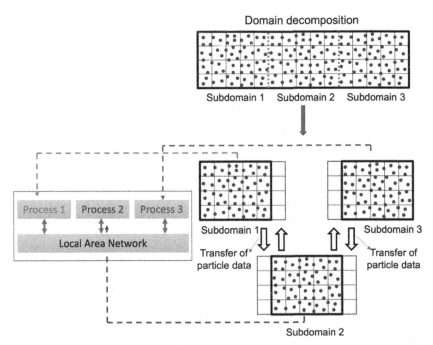

Figure 9.5 The division of subdomains and communications between subdomains occurred at the boundary of subdomains [1].

the subdomain boundary is calculated, some neighbor particles may locate in the adjacent subdomain. Therefore communications between adjacent subdomains (processes) are indispensable. As shown in Fig. 9.5, only the particles located in the background grid which are closest to the subdomain boundary should be transferred to the adjacent subdomains. Before a particle interacts with its neighbor particles, particle information should be transferred to adjacent subdomains.

The load balance is also an important issue when the program is accelerated by MPI. To solve this problem the subdomains' boundaries can be adjusted dynamically by comparing the particle number of the current subdomain and the neighbor subdomain [2]. As shown in Fig. 9.6A, when the particle number of the left subdomain is obviously less than that of the right subdomain, the domain boundary will be moved to the right, and then the particles between the old and new boundaries will be transferred to the left subdomain. After the adjusting process the particle number of the two subdomains is almost identical as shown by Fig. 9.6B. After these operations, all the domain boundaries can be dynamically adjusted to keep load balanced.

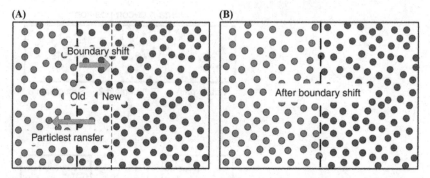

Figure 9.6 The process of the redivision of the subdomain. When the left particle number is less than that of the right subdomain, move the dividing boundary to the right and transfer the particles between the old and new boundaries to the left subdomain [1]. (A) Before adjusting. (B) After adjusting.

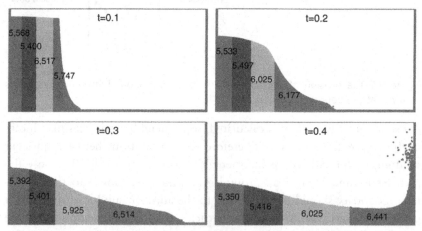

Figure 9.7 The profiles of particles assigned to different processes at four moments are shown. The particles of different colors in the figures are calculated by different processes. The number of particles assigned to each process is also shown in the figure [1].

After the dynamic movement of the internal boundaries the number of particles in each subdomain are approximately identical. An example of dynamic load balance is shown in Fig. 9.7, as presented in Ref. [1].

Clusters are rather popular architectures of the high-performance computers. In a cluster, nodes are connected to each other with a high-speed local area network as shown in Fig. 9.8. Most nodes in a cluster are multicore computers. Therefore the combination of the OpenMP and MPI

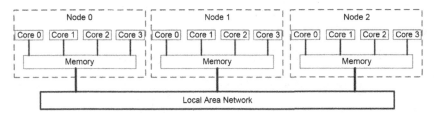

Figure 9.8 Computer clusters where multicore nodes are connected by local area networks [1].

models is of great potential to accelerate the MPS method on a cluster. First the calculation domain is divided into several subdomains. These subdomains are assigned to different processes which are executed on different nodes, and the MPI model is used to exchange information among these nodes. Second the process on each node forks several threads which are assigned to different cores of the node to calculate the motion of the particles concurrently. For example the OpenMP model is used to accelerate the program on each node. The hybrid OpenMP/MPI method is especially suitable for a cluster as shown in Fig. 9.8.

Because the MPS method is a semi-implicit method, an easy-to-parallelize solver must be selected to parallelize the whole program. The comparison of different solvers for semi-implicit particle methods was investigated by Duan and Chen [1]. When the original symmetric Laplacian model is used to discretize the pressure Poisson equation (PPE), the Conjugate Gradient (CG) solver [4] is the best choice in terms of simplicity, stability, and parallel efficiency. When the high-order asymmetric Laplacian model is employed to discretize PPE, the Bi-Conjugate Gradient Stabilized (BiCGStab) solver [4] seems the best choice. When a multiphase flow with a high density ratio is simulated, the Jacobi (or diagonal) preconditioned BiCGStab solver [4] is the best choice based on our experiences.

The domain decomposition and dynamic load balance are the most difficult parts in the MPI parallelization. To achieve a high parallel efficiency, one must make sure that (1) each subdomain almost calculates the same number of particles and (2) the intersection area of all subdomains should be the minimized to reduce the communication among different processes. To achieve the two targets the MPI algorithm usually becomes quite complicated. Fortunately, there are already some open-source parallel libraries which can be directly employed in the particle methods. To our best knowledge the following libraries/programs are recommended to employ or to refer to when the readers want to develop a high–efficiency parallel program for their particle methods.

1. LexADV_EMPS (c): https://adventure.sys.t.u-tokyo.ac.jp/lexadv/lexadv_EMPS.html

2. PPM (Fortran): http://www.ppm-library.org

3. LAMMPS (c++): https://www.lammps.org

4. SPHinXsys (c++): https://xiangyu-hu.github.io/SPHinXsys/

5. DualSPHysics (c++): https://dual.sphysics.org

6. PHANTOM (Fortran): https://phantomsph.bitbucket.io

9.3 Graphics processing unit acceleration

Most of the available parallel programs developed in the past several decades are generally based on the MPI strategy, which commonly requires strong supports from costly workstations or supercomputers. Graphics processing units (GPUs), which are new architectures for massive parallel processing and computer clusters, recently became widespread. Although they were developed exclusively for graphical purposes and driven by the computer game industry, the development of GPU technology is leading to the development of general-purpose GPUs (GPGPUs) because floating-point arithmetic by GPUs became possible and the compute unified device architecture (CUDA) along with its software development kit (SDK) has been released.

Hori et al. [5] developed a GPU-accelerated version of the standard MPS code. A two-dimensional calculation of elliptical drop evolution with 100k particles showed that the GPU-accelerated version was approximately 7 times faster than the code for a CPU using only one core. Zhu et al. [6] developed a GPU-based MPS model with four optimization levels. The efficient neighborhood particle search is performed using an indirect method. A simulation of a benchmark problem of dam-breaking flow showed that the GPU-based MPS model outperforms the traditional CPU model by 26 times. Kakuda et al. [7] presented a GPU-based MPS method for calculating 2-D and 3-D dam break problems. Compared with CPU simulations, the speedups are 12 times and 17 times, respectively. Gou et al. [8] simulated the isothermal multiphase fuel–coolant interaction by using the MPS method with GPU acceleration. These results show that with the help of GPUs, speedups about 1 order of magnitude could be demonstrated.

The CUDA programming model is illustrated in Fig. 9.9. A CUDA program is divided into a host part and a device part. The host part runs on CPU, while the device part runs on GPU. Functions executed on GPU are named kernel functions, which are allowed to be called by CPU. For programmers, only two things need to be done while they are programming on GPU. First, before the use of the parallel power of GPU, it is very important to make sure that the algorithm of kernel functions is suitable for parallel computing. Fortunately, through some simple modifications, most of the algorithms can be parallelized. Second the programmer is required to divide the data into many blocks and decide how many threads should be used in each block. After the above two steps the developed kernel functions can run in parallel on GPU automatically. With CUDA, programmers can use the maximum potential capacity of

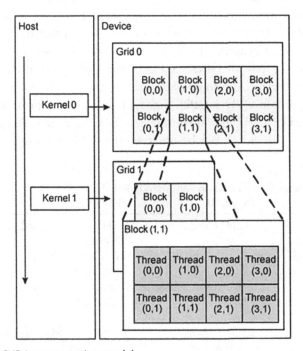

Figure 9.9 CUDA programming model.

the GPU in a convenient manner. The Link-list calculation can be separated in four steps in CUDA. In step 1, background cells are generated. In step 2, particle cell index is found by the position of the particle. In step 3, particles are sorted by the particle cell index. In step 4 the start particle of each cell is found. In step 5, neighboring particles are searched from these cells. In this way, particles in the same cell are stored consecutively and coalescent memory access is achieved while particle interaction is calculated for neighboring particles.

In Ref. [5] the results of simulations by using both the CUDA-MPS calculation code and the conventional sequential calculation code are discussed. Computational environments are shown in Table 9.1. The dam break problem with different particle resolutions is calculated. The total numbers of particles are 7080, 11676, 24360, 89520, and 109211 in each case by changing the diameter of the particle to 4.0, 3.0, 2.0, 1.0, and 0.9 mm, respectively.

Fig. 9.10 shows the time fraction in the five cases simulated by only CPU or GPU + CPU. Each segment shows, beginning, search for neighboring particles (neighboring), generation of matrix (matrix), preconditioning by diagonal scaling (scaling), iterative computation (pcg solver), external calculation (gravity and viscosity term external), writing output into text files (output), and others. It must be noted that the part of output includes time dedicated to the transfer of memory between the device and host in the case of GPU + CPU. From this figure, it can be stated that the iterative calculation takes up most of the calculation time both on only CPU and on GPU + CPU. Fig. 9.11 shows the improvement factor of using GPU + CPU in comparison to only CPU results. Regarding relatively many particles, it can be observed that GPU + CPU is more than

Table 9.1 Computational environment used in Ref. [5].

	Only CPU	GPU + CPU
CPU	Intel" Core i7 920 2.G7 GHz (1 core is used in this paper)	
Main memory	2.49 GB	
Graphics card	NVIDIA Tesla C 1060 4 GB 1 & 5.S5	
Driver version	185.85	
OS	Microsoft Windows XP Professional Version 2002 SP3	
Programming language	Fortran	CUDA, C
Compiler	Intel" Fortran Compiler 9.1	CUDA 2.1, Microsoft Visual Studio 2005

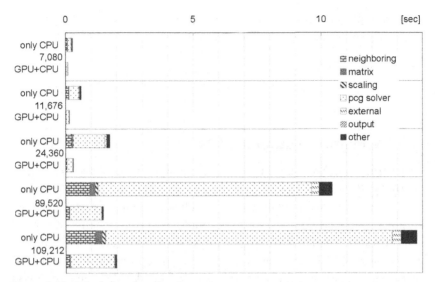

Figure 9.10 Calculation time fraction.

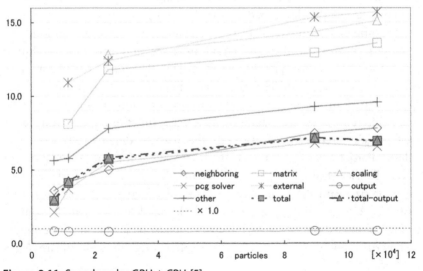

Figure 9.11 Speedups by GPU + CPU [5].

10 times faster than only CPU in generation of matrix, preconditioning, and external calculations. In the kernels handling these tasks, one thread can process one particle more easily than in the other kernels.

Nevertheless, the use of a single-GPU card is not sufficient for engineering applications that require several million particles to predict the desired physical processes as the available memory space is insufficient. It is essential to harness the performance of multiple GPUs for large simulations.

9.4 Multiresolution models

Particle methods are different from mesh methods which can easily use the grids with different sizes at different regions. Several attempts have been tried to use multiresolution scheme in single-particle hydrodynamics (SPH) method. However, multiresolution scheme for MPS method is more difficult to implement because the PPE is solved implicitly. Traditional techniques of keeping the particle number density (PND) constant and detecting surface particles based on the deficiency of PND cannot be applied anymore. In recent years, some noteworthy developments are reported for MPS method to use different spatial resolutions.

Yoon et al. [9] developed a particle multiple-resolution model to a particle-grid hybrid method. In the simulation of bubble generation, fine resolution is used at the gas–liquid interface and this fine resolution follows bubble motion [10]. Chen et al. [11] developed particle interaction models among particles that have different sizes by using a cubic spline kernel function to calculate the contributions from the particles with different sizes. An additional weight function is introduced as well for clustered particles to keep the particle distances at reasonable values. Particle volumes are defined and used for the calculation of PND. Tanaka et al. [12] defined different effective radii for particles with different sizes and used the average value of effective radii between two particles. A parameter of packing ratio calculated by the particle volume to the volume in the effective domain is used to replace PND. The modified particle interaction models are implemented in least square MPS method.

The common feature of various particle interaction models between particles with different sizes was summarized by Koshizuka et al. [13]. The volume of each particle i V_i is defined as the particle length L_i to the power of the number of dimensions d:

$$V_i = L_i^d \tag{9.1}$$

The PND, gradient, and Laplacian models are expressed with the volume of each particle as follows:

$$n_i = \sum_{j\neq i} \frac{V_j w\left(|\mathbf{r}_j - \mathbf{r}_i|, r_{ei}\right) + V_i w\left(|\mathbf{r}_j - \mathbf{r}_i|, r_{ej}\right)}{2V_i} \tag{9.2}$$

$$\langle\phi\rangle_i = \frac{d}{V_i n^0} \sum_{j\neq i} \left[\frac{\phi_j - \phi_i}{|\mathbf{r}_j - \mathbf{r}_i|^2} (\mathbf{r}_j - \mathbf{r}_i)\, V_j w\left(|\mathbf{r}_j - \mathbf{r}_i|, r_{ei}\right) \right] \tag{9.3}$$

$$\langle\nabla^2\phi\rangle_i = \frac{2d}{V_i \Lambda_i} \sum_{j\neq i} \left[(\phi_j - \phi_i) \frac{V_j w\left(|\mathbf{r}_j - \mathbf{r}_i|, r_{ei}\right) + V_i w\left(|\mathbf{r}_j - \mathbf{r}_i|, r_{ej}\right)}{2} \right] \tag{9.4}$$

where r_{ei} and r_{ej} are the effective radii of particles i and j, respectively, which are related to the sizes of particles i and j. The scalar parameter Λ_i is calculated by

$$\Lambda_i = \sum_{j\neq i} \left[|\mathbf{r}_j - \mathbf{r}_i|^2 \frac{V_j w\left(|\mathbf{r}_j - \mathbf{r}_i|, r_{ei}\right) + V_i w\left(|\mathbf{r}_j - \mathbf{r}_i|, r_{ej}\right)}{2V_i} \right] \tag{9.5}$$

The particle splitting and merging procedures proposed by Takana et al. [14] are introduced in the following part.

To conserve the particle volume the particle size after splitting one particle into two particles is calculated by

$$L_{i'} = \left(\frac{1}{2}\right)^{\frac{1}{d}} L_i \tag{9.6}$$

The positions of new particles after splitting are given by

$$\mathbf{r}_{i'} = \mathbf{r}_i \pm \frac{\alpha_t}{2} L_{i'} \mathbf{t}_i \tag{9.7}$$

where $\mathbf{t}$ is a direction vector along which particles are going to be divided; the subscript i' indicates the new particle. The direction vector $\mathbf{t}$ should be determined by minimizing the error function that is defined by

$$e_i = \max\left(e_{ij}\right)$$

$$e_{ij} = \begin{cases} \dfrac{L_{ij} - r_{i'j}}{L_{ij}} & \left(r_{i'j} < L_{ij}\right) \\[2mm] 0 & \left(r_{i'j} \geq L_{ij}\right) \end{cases} \tag{9.8}$$

where $L_{ij} = \frac{L_i + L_j}{2}$ is the average diameter between particles i and j. If the error e_i at a splitting direction can be smaller than the threshold value α_e, the splitting proceeds; otherwise the splitting process stops. The splitting of the free-surface particle is performed at the direction perpendicular to the normal vector of free surface.

The velocities of the new particles after splitting are calculated by the following interpolation equation:

$$\mathbf{u}_{i'} = \mathbf{u}_i + \langle \nabla \mathbf{u} \rangle_i (\mathbf{r}_{i'} - \mathbf{r}_i) \tag{9.9}$$

In the particle merging process the position and velocity of the newly generated particle are calculated as below by conserving the angular and linear momentums:

$$\mathbf{r}_{i'} = \frac{V_i \mathbf{r}_i + V_j \mathbf{r}_j}{V_i + V_j} \tag{9.10}$$

$$\mathbf{u}_{i'} = \frac{V_i \mathbf{u}_i + V_j \mathbf{u}_j}{V_i + V_j} \tag{9.11}$$

The diameter of the new particle after merging is evaluated by considering volume conservation, that is,

$$L_{i'} = \left(L_i^d + L_j^d \right)^{\frac{1}{d}} \tag{9.12}$$

In the multiresolution scheme of Ref. [12], to keep the accuracy and stability, one particle is allowed to split into a maximum of two particles with the conservation of particle volume and vice versa. The particle sizes before and after splitting and merging should not have a large difference, which are judged using a criterion. The multiresolution scheme is verified by the simulations of Poiseuille flow, multidiameter channel flow, Taylor green vortex, and the flow in gears.

Chen et al. [11] proposed a multistep scheme for particle splitting and coalescing. One particle splits into seven daughter particles by five steps (see Fig. 9.12), which enables the daughter particles to find their proper

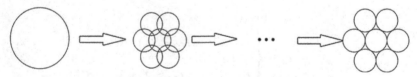

Figure 9.12 Large particle splitting into daughter particles by five steps [11].

positions. The daughter particles have the same translational velocities as the mother particle for the conservation of translational momentum but have no angular velocities because the mother particle has no angular velocity. Similarly the particle merging process completes by five steps as well, but it only takes place between two small particles. The velocity and position of the new particle after merging are calculated using the same scheme as that of Tanaka et al. [14] by conserving the volume and momentum.

Khayyer et al. [15] applied multiresolution scheme to incompressible fluid—elastic structure interactions, where the fluid and structure are discretized with coarse and fine particles, respectively. The effective radii of all particles are set as a constant value regardless of particle scales to ensure the consistency in force calculation of particle pairs. The space potential particles are introduced to deal with the void spaces formed by particles with various sizes. The formulas of weight function and PND are modified, and the concept of PND is used to enhance volume conservation at the fluid—structure interface to ensure consistency of particle-based discretization and to enhance the imposition of boundary conditions.

The above-mentioned multiresolution scheme is usually applied to single-phase flow, and the region interface of fine and coarse particles is fixed. To make the multiresolution scheme more applicable an adaptive multiresolution scheme for multiphase flow is proposed by Liu and Zhang [16]. Particle resolution is adaptively changed by judging the distance from the interface. The smallest particles are used to the layer nearest to the interface, and the largest particles are used to layer farthest from the interface. The particle velocities and positions after splitting and merging are calculated by interpolation method. The capability of the adaptive multiresolution scheme is demonstrated by using both two-layer and three-layer adaptive refinements in the simulations of static two-fluid pool, Rayleigh—Taylor instability, and single rising bubble.

Shibata et al. [17] developed an ellipsoidal particle model to reduce the computational cost. The particle distances at x and y directions are different. A coordinate transformation scheme is proposed to treat ellipsoidal particles in the same manner as sphere particles. The coordinate transformation is conducted as follows:

$$\hat{x}_\alpha = C_{\alpha\beta}^{-1} x_\beta \qquad (9.13)$$

$$\hat{x}_\beta = C_{\alpha\beta}^{-1} x_\alpha \qquad (9.14)$$

where parameter x is a coordinate in the physical coordinate system, $\hat{x}$ is the coordinate in the system after the coordinate transformation, the subscripts α and β denote the directions, and $C_{\alpha\beta}$ is a component of a matrix C, which is expressed as

$$\mathbf{C} = \begin{pmatrix} C_{xx} & C_{xy} & C_{xz} \\ C_{yx} & C_{yy} & C_{yz} \\ C_{zx} & C_{zy} & C_{zz} \end{pmatrix} = \begin{pmatrix} s_x & 0 & 0 \\ 0 & s_y & 0 \\ 0 & 0 & s_z \end{pmatrix} \tag{9.15}$$

where the parameters s_x, s_y, s_z are the real aspect ratios of each particle in the physical coordinate system. In the coordinate system of $\hat{x}$ the particle scales are the same at all directions, indicating that the ellipsoidal particles can be treated as spherical particles.

Therefore the Navier–Stokes equation and the PPE in the transformed coordinate system can be expressed as

$$\rho^0 \frac{Du_\alpha}{Dt} = -C_{\alpha\beta}^{-1} \frac{\partial P}{\partial \hat{x}_\beta} + \mu \left(C_{\beta\beta}^{-1} \right)^2 \frac{\partial^2 u_\alpha}{\partial \hat{x}_\beta \partial \hat{x}_\beta} + \rho^0 g_\alpha \tag{9.16}$$

$$\left(C_{\beta\beta}^{-1} \right)^2 \frac{\partial^2 P}{\partial \hat{x}_\beta \partial \hat{x}_\beta} = -\frac{\rho^0}{\Delta t^2} \frac{\hat{n}^* - \hat{n}^0}{\hat{n}^0} \tag{9.17}$$

where a coefficient C^{-1} is applied to the spatial differentials. The PND and gradient model are calculated based on the coordinates after transformation.

The Laplacian model in the original MPS method is modified as follows to make it applicable to ellipsoidal particles:

$$\langle \nabla^2 \phi \rangle_i^k = \left\langle \frac{\partial^2 \phi}{\partial x_\beta \partial x_\beta} \right\rangle_i^k = \left\langle \left(C_{\beta\beta}^{-1} \right)^2 \frac{\partial^2 \phi}{\partial \hat{x}_\beta \partial \hat{x}_\beta} \right\rangle_i^k$$

$$= \frac{2d}{\hat{n}^0} \sum_{j \neq i} \left\{ \frac{1}{\lambda_x^0} \frac{\left(x_j^k - x_i^k \right)^2}{\left| \vec{r}_j^k - \vec{r}_i^k \right|^2} + \frac{1}{\lambda_y^0} \frac{\left(y_j^k - y_i^k \right)^2}{\left| \vec{r}_j^k - \vec{r}_i^k \right|^2} \right.$$

$$\left. + \frac{1}{\lambda_z^0} \frac{\left(z_j^k - z_i^k \right)^2}{\left| \vec{r}_j^k - \vec{r}_i^k \right|^2} \right\} \left(\phi_j^k - \phi_i^k \right) w \left(\left| \vec{r}_j^k - \vec{r}_i^k \right| \right) \tag{9.18}$$

where λ_x^0, λ_y^0, and λ_z^0 are the scalar parameters used to calibrate Laplacian model, which are calculated in the same manner as the standard MPS method.

Shibata et al. [18] developed particle overlapping technique for multi-resolution simulation of particle method. The computation domain is divided into several subdomains, and the adjacent subdomains partially overlapped with each other. Different particle sizes are employed for every domain, and the pressure and velocity information is interpolated at the overlapping regions of subdomains, as illustrated in Fig. 9.13. The particles only interact with other particles in the same subdomain, and the calculation results at one time step are assigned to other sub-domains as a boundary condition of the inlet/outlet boundary. Several subregions are divided within the overlapping region, and their lengths are defined as $L_{\text{margine,coarse}} = 6l_{0,\text{coarse}}$, $L_{\text{margine,fine}} = 6l_{0,\text{fine}}$, $L_{\text{fix,fine}} = 4l_{0,\text{fine}}$, and $L_{\text{fix,coarse}} = 4l_{0,\text{coarse}}$. Shibata et al. [18] used the moving least squares method to interpolate velocity and pressure between the coarse and fine particles in the overlapping region, but other interpolation methods are also applicable.

The computational procedure of the overlapping particle technique is described as follows based on Fig. 9.13. In each time step the subdomain of coarse particles is calculated first, but it needs the temporary velocities of fine particles that locate in the region near the inlet/outlet boundary of the coarse particle subdomain. The temporary velocity in the region of fine particle subdomain that has a distance of $L_{\text{overlaping}} + r_{e,\text{coarse}} + l_{0,\text{coarse}}$ from the outlet boundary of fine particle subdomain is calculated by

$$\vec{u}^*_{\text{fine}} = \left(\nu \nabla^2 \vec{u}^*_{\text{fine}} + \vec{g} \right) \Delta t_{\text{coarse}} + \left(\vec{u}^*_{\text{fine}} \right) \tag{9.19}$$

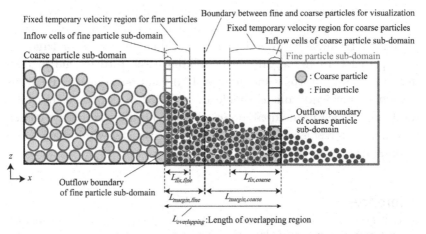

Figure 9.13 Conceptual diagram of the bilateral overlapping technique [18].

The calculated velocity $\vec{u}^*_{\text{fine}}$ is only used for interpolation of the velocity of coarse particles located in the "fixed temporary velocity region for coarse particles," and the interpolated velocity of coarse particle serves as the boundary condition of the inlet/outlet boundary. Similarly the pressure of fine particles at the previous time step is interpolated to the coarse particles on the boundary and then serves as the boundary condition.

After the calculation of coarse particles in one time step the calculation of fine particles is carried out. The temporary velocities of coarse particles in the region of coarse particle subdomain with a distance of $L_{\text{overlaping}} + r_{e,\text{fine}} + l_{0,\text{fine}}$ from the outlet boundary of coarse particle subdomain are calculated by

$$\vec{u}^*_{\text{coarse}} = \left(\nu \nabla^2 \vec{u}^K_{\text{coarse}} + \vec{g}\right)\Delta t_{\text{fine}} + \left(\vec{u}^K_{\text{coarse}}\right) \tag{9.20}$$

The $\vec{u}^*_{\text{coarse}}$ is also only used to velocity interpolation of fine particles in the region of "fixed temporary velocity region for fine particles," and the interpolated velocities of fine particles serves as boundary condition in the calculation of fine particles. Different scales of time step can be used for the calculations of coarse and fine particles.

The overlapping particle technique can be used to various particle methods, and it can avoid the error and instability caused by the interactions of particles of different sizes because the particle size is uniform in one subdomain.

In a summary of the above studies, multiresolution technique has not got sufficient development in MPS method, but it is an effective approach to reduce computation cost. The difficulties lie in the PND calculation that varies according to the scales of neighboring particles and the particle volume that cannot be considered as constant. Particle interaction models should be revisited by considering the volume difference of large and small particles. In the algorithms of splitting large particles and merging small particles the focus is the jump of PND and thus the effective radius of particle interaction model should be averaged or smoothed. Instability is still a challenge in multiresolution scheme, and the consistency, conservation, and convergence in the interaction of particles with different sizes are important research topics.

References

[1] G.T. Duan, B. Chen, Comparison of parallel solvers for moving particle semi-implicit method, Eng. Comput. 32 (2015) 834–862.

[2] B.D. Rogers, R.A. Dalrymple, P.K. Stansby, D.L. Laurence, Development of a Parallel SPH code for free-surface wave hydrodynamics, in: Proceedings of 2nd International SPHERIC Workshop, Madrid, 2007, pp. 111−114.

[3] D.W. Holmes, J.R. Williams, P. Tilke, A framework for parallel computational physics algorithms on multi-core: SPH in parallel, Adv. Eng. Softw. 42 (2011) 999−1008.

[4] T.F. Chan, [Book Review] Iterative Methods for Sparse Linear Systems, Society for Industrial Mathematics, Philadelphia, 2003.

[5] C. Hori, H. Gotoh, H. Ikari, A. Khayyer, GPU-acceleration for moving particle semi-implicit method - ScienceDirect, Comput. Fluids 51 (1) (2011) 174−183.

[6] X.S. Zhu, L. Cheng, L. Lu, B. Teng, Implementation of the moving particle semi-implicit method on GPU, Sci. China 54 (3) (2011) 523−532.

[7] K. Kakuda, T. Nagashima, Y. Hayashi, S. Obara, J. Toyotani, S. Miura, et al., Three-dimensional fluid flow simulations using GPU-based particle method, Comput. Model. Eng. Sci. 93 (5) (2013) 363−376.

[8] W. Gou, S. Zhang, Y. Zheng, Simulation of isothermal multi-phase fuel-coolant interaction using MPS method with GPU acceleration, Kerntechnik 81 (3) (2016) 330−336.

[9] H.Y. Yoon, S. Koshizuka, Y. Oka, A particle-gridless hybrid method for incompressible flows, Int. J. Numer. Methods Fluids 30 (1999) 407−424.

[10] H.Y. Yoon, S. Koshizuka, Y. Oka, Direct calculation of bubble growth, departure, and rise in nucleate pool boiling, Int. J. Multiph. Flow. 27 (2001) 277−298.

[11] X. Chen, Z. Sun, L. Liu, G. Xi, Improved MPS method with variable-size particles, Int. J. Numer. Methods Fluids 80 (2016) 358−374.

[12] M. Tanaka, R. Cardoso, H. Bahai, Multi-resolution MPS method, J. Comput. Phys. 359 (2018) 106−136.

[13] S. Koshizuka, K. Shibata, M. Kondo, T. Matsunaga, Moveing Particle Semi-Implicit Method: A Meshfree Particle Method for Fluid Dynamics, Acdemic Press, 2018.

[14] M. Takana, T. Masunaga, Y. Nakagawa, Multi-reolution MPS method, Trans. JSCES (2009).

[15] A. Khayyer, N. Tsuruta, Y. Shimizu, H. Gotoh, Multi-resolution MPS for incompressible fluid-elastic structure interactions in ocean engineering, Appl. Ocean. Res. 82 (2019) 397−414.

[16] X. Liu, S. Zhang, Development of adaptive multi-resolution MPS method for multiphase flow simulation, Comput. Methods Appl. Mech. Eng. 387 (2021) 114184.

[17] K. Shibata, S. Koshizuka, I. Masaie, Cost reduction of particle simulations by an ellipsoidal particle model, Comput. Methods Appl. Mech. Eng. 307 (2016) 411−450.

[18] K. Shibata, S. Koshizuka, T. Matsunaga, I. Masaie, The overlapping particle technique for multi-resolution simulation of particle methods, Comput. Methods Appl. Mech. Eng. 325 (2017) 434−462.

CHAPTER TEN

Applications in nuclear engineering

Owing to the advantages of moving particle semi-implicit (MPS) method in simulating multiphase and multicomponent flows, it has been widely applied to nuclear engineering in recent years. In the scope of fundamental thermal hydraulics, the bubble dynamics in the evaporation and condensation processes are analyzed with MPS method. In terms of nuclear reactor severe accidents, the phenomena of fuel rod disintegration and melt behaviors in the lower head and at the containment are investigated by coupling advanced models into MPS method.

10.1 Fundamental thermal hydraulics

Many studies are performed to investigate the bubble dynamics in nuclear reactor thermal hydraulics, including the boiling bubble growth, bubble condensation, bubble coalescence, and void bubble rising in liquids. Yoon et al. [1] simulated bubble growth, departure, and deformation processes in nucleate pool boiling using MPS coupled with meshless advection using flow-directional local grid (MPS-MAFL) method, as shown in Fig. 10.1. The liquid phase is modeled using moving particles, but the bubble is not represented with particles; the interface of the liquid and bubble is tracked by the movable positions of interfacial liquid particles. The superheated liquid microlayer beneath the bubble is not modeled, but its effect is placed over the vapor bubble at the beginning of bubble formation. The analysis results indicate that bubble departure diameter is in proportion to contact angle and the square root of surface tension. The simulated bubble growth rate, departure radius, and heat transfer rate show good agreements with the experimental results. Heo et al. [2] investigated bubble growth in transient pool boiling with high heat flux and high subcooling conditions using the MSP-MAFL method. It is found that initial bubble radius has little effect on growth initiation

Moving Particle Semi-implicit Method.
DOI: https://doi.org/10.1016/B978-0-443-13508-8.00010-X

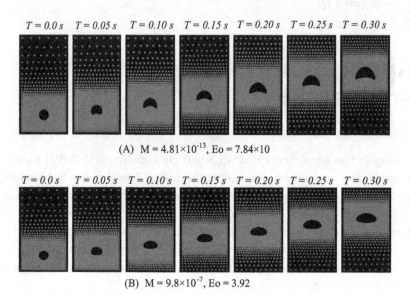

Figure 10.1 Numerical simulation of a rising bubble [1].

time and bubble departure radius. Chen et al. [3] investigated the effect of bulk liquid velocity, liquid subcooling, wall superheat, and surface incline angle on bubble dynamics including bubble shape, liftoff diameter, and liftoff time in flow boiling conditions. Tian et al. [4] applied MPS-MAFL method to the simulation of bubble condensation in water. The coupling process of momentum and energy equations are specially treated, and the heat flux at the interface is calculated according to the energy variation of the interfacial liquid layer. Bubble deformation, lifetime, and bubble size history under lower and higher degrees of liquid subcooling and different system pressures are analyzed. Chen et al. [5] simulated coalescence dynamics of a pair of bubbles rising in a stagnant liquid, as presented by Fig. 10.2. The shape variation and velocity and pressure fields of the leading and the tail bubbles are studied. Li et al. [6] and Zuo et al. [7] simulated argon and nitrogen bubble rising behaviors in liquid metals, which extend MPS method to the thermal hydraulics research of advanced reactors. Chen et al. [8] reviewed the achievements on bubble dynamics analysis using MPS method.

Regarding the gas—liquid two-phase flow in the fuel bundle region, Xie et al. [9] investigated liquid droplet deposition in the annular-mist flow regime of a boiling water reactor (BWR) to analyze the possibility and the effect of splash occurrence. Guo et al. [10] simulated droplet

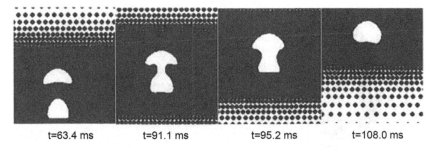

t=63.4 ms	t=91.1 ms	t=95.2 ms	t=108.0 ms

Figure 10.2 Coalescence of a pair of bubbles rising in a stagnant liquid [5].

entrainment in annular two-phase flow by a hybrid method, in which the gas and liquid phases are discretized by grid and particle separately. Shirakawa et al. [11,12] studied the flows around a BWR spacer, sub-cooled boiling with special interest in the location of void departure, and the void distribution in a circular tube by using a two-fluid particle inter-action model in the framework of MPS method.

On the other aspects of nuclear reactor thermal hydraulics, some simu-lations are also conducted with MPS method. Ui et al. [13] developed a solid−liquid multiphase model to simulate debris transport, settling, and resuspension in the case of a loss-of-coolant accident of a pressurized water reactor (PWR). The model can solve the motion of solid particles of different sizes. Turbulence model is also coupled into MPS method. With the improved method, the analysis of debris transport process in a full-scale PWR containment vessel floor is carried out. Xiong et al. [14] simulated droplet impingement onto a rigid wall in two and three dimen-sions, which could lead to pipe wall thinning. The impact pressure and shear stress distributions are sensitively evaluated under different droplet diameters and impact obliqueness. The effect of water film on droplet impingement damage is investigated as well. The comparison with dry wall conditions indicates that water film can effectively mitigate or even eliminate high peak pressure that appears at the contact edge [15]. Wang et al. [16,17] simulated water-flooding processes in a nuclear reactor building of AP1000 with explicit MPS (EMPS) method, as shown in Fig. 10.3, in which multiple floating bodies are considered. The solid−−fluid and solid−solid interactions are modeled by passively moving solid (PMS) model and discrete element method (DEM), respectively. The capability for large-scale simulation on nuclear thermal hydraulics by EMPS method is proved.

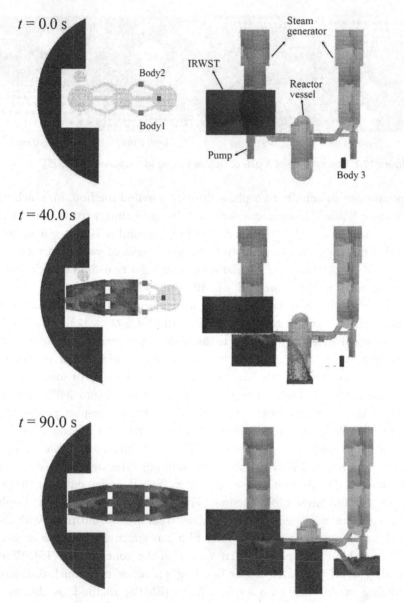

Figure 10.3 Flooding of the nuclear reactor building with multiple floating bodies [17].

10.2 Melt behavior in severe accidents

10.2.1 Fuel rod disintegration

A COMPASS (Computer Code with Moving Particle Semi-implicit for Reactor Safety Analysis) code was developed by Japan universities and institutes on the basis of MPS method to study the key phenomena in core disruptive accidents of sodium–cooled fast reactors. The fuel pellet cracking and cladding disruption due to high pressure that is built up by fuel vapor and released fission product gas are investigated by coupling the thermofluid and structural mechanics. Figs. 10.4 and 10.5 present the cracking behavior of the fuel pellet and the disruption behavior of the fuel rod. The phenomena are consistent with our empirical knowledge on fuel rod rupture propagation, but the detailed validations have not be performed [18].

Eutectic reaction is an important phenomenon occurring at the early stage of a nuclear reactor severe accident. The materials may liquefy at a temperature below their melting points. In the light water reactors, the material of fuel rod cladding is usually zirconium alloy, and at high temperature and in a steam environment the outermost layer of zirconium alloy can be oxidized to zirconia. Eutectic reaction may take place among the materials of zirconium alloy, zirconia, uranium dioxide, and stainless-steel structures. Eutectic reaction is a microphenomenon arising from the atomic diffusion. Many binary and ternary phase diagrams have been produced from the material experiments, which are important references for severe accident analysis. As the temperature further increases after the material eutectic reaction, the cladding material of the fuel rod melts and migrates downward.

According to the phase diagram of materials and based on mass diffusion theory, the eutectic reaction model is first developed in COMPASS code. Later, it is improved and applied to the melt behavior analysis of light water reactors by Mustari and Oka [19,20]. The mass diffusion process is calculated according to Fick's second law expressed as

$$\frac{\partial m}{\partial t} = D\nabla^2 m \tag{10.1}$$

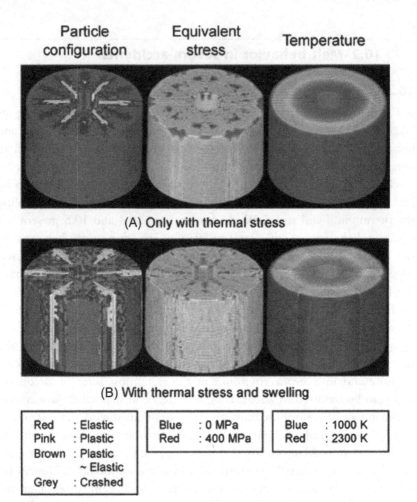

Red	: Elastic		Blue	: 0 MPa		Blue	: 1000 K
Pink	: Plastic		Red	: 400 MPa		Red	: 2300 K
Brown	: Plastic						
	~ Elastic						
Grey	: Crashed						

Figure 10.4 Cracking behavior of the fuel pellet [18]. (A) Only with thermal stress. (B) With thermal stress and swelling.

where m is the mass and D is the diffusion coefficient of the material. Discretizing Eq. (10.1) with the Laplacian model of MPS method, the mass diffused into particle i can be calculated as

$$m_i^{t+1} = m_i^t + D\Delta t \cdot \frac{2d}{n^0 \lambda} \sum_j (m_j^t - m_i^t) w_{ij} \qquad (10.2)$$

The model is applied to the analyses of eutectic reactions of U–Fe solid–liquid system and Pb–Sn solid–solid system. Fig. 10.6 shows the

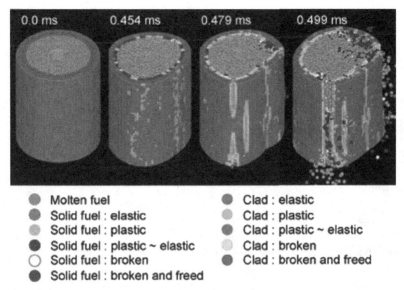

● Molten fuel   ● Clad : elastic
● Solid fuel : elastic   ● Clad : plastic
● Solid fuel : plastic   ● Clad : plastic ~ elastic
● Solid fuel : plastic ~ elastic   ● Clad : broken
○ Solid fuel : broken   ● Clad : broken and freed
● Solid fuel : broken and freed

Figure 10.5 Disruption behavior of the fuel rod [18].

penetration rate of stainless steel 304 by uranium. The analyzed results agree well with the experimental data.

Li et al. [21] investigated the dissolution process of uranium dioxide by molten zircaloy using mass diffusion model. The results indicate that the dissolution of UO_2 by molten zircaloy fits the parabolic law, but the dissolution rate decreases after UO_2 reaches saturation in liquids. Natural convection of molten zircaloy has a significant effect on the dissolution process. In order to improve the calculation accuracy of mass and heat transfer at the phase interface, Li et al. [22] introduced meshes to the calculations. The meshes are only generated for boundary layers by connecting the particle centers, whereas the mass transfer and heat transfer of inner particles are still calculated by the particle interaction models. Computation cost increases significantly as more particles participate into mesh generation, especially for three dimensions. Based on this coupled method, the cladding oxidation (see Fig. 10.7) and dissolution behavior between ZrO_2 and molten zirconium are analyzed. Wang et al. [23] applied MPS method to the simulations of melt migration and solidification phenomena in the fuel rod channel of the lead cooled reactor. An initial melt block is assumed, and it moves upward under the effect of buoyancy. The effect of initial solid fraction on melt redistribution is investigated, as shown in Fig. 10.8, where the initial solid fraction is

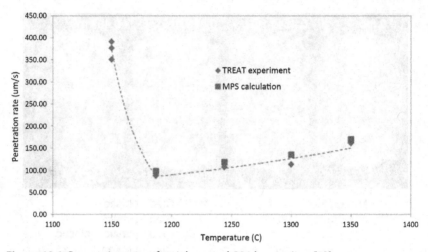

Figure 10.6 Penetration rate of stainless steel 304 by uranium [19].

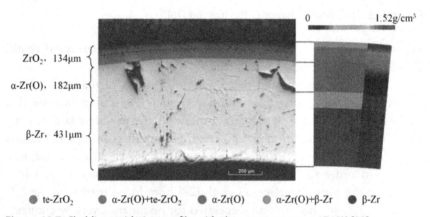

Figure 10.7 Cladding oxidation profile with the temperature at 1748K [22].

0.5465. When the melt migrates from the core region to the lower plenum, the flow channels in fuel support pieces of a BWR are the potential relocation channels. Melt solidification behavior in fuel support pieces can affect melt relocation process and determine whether the melt pool may form above the core plate. The analysis by Chen et al. [24] points out that the channels in fuel support pieces are plugged due to the solidification of zircaloy melt when the initial temperature is low, and fuel support pieces are intact in all the calculated cases.

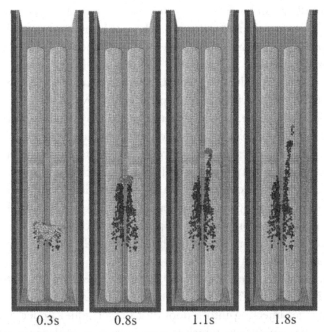

0.3s   0.8s   1.1s   1.8s

Figure 10.8 Melt solidification process in the fuel rod channel [23].

10.2.2 Melt behavior in RPV lower head

Following the disintegration of fuel rod bundles, the melt and fragments slump into the lower head where the residual water may exist. During the interaction of melt with water (also named fuel—coolant interaction), the melt jet first breaks up into many droplets under the impact of hydraulic resistance. Violent heat transfer and water evaporation phenomena present at the melt and coolant interface. With the rapid increase of steam volume, a pressure pulse is produced and expands in the coolant. The pressure pulse drives the melt droplet further to breakup into many small fragments. The fragment size distribution and contact area with the coolant are important parameters affecting the heat transfer and melt coolability. With the coolant boiling off a debris bed forms comprising metallic and oxidic materials. The debris will melt again as the decay heat continuously releases, and the melt migrates in the void space. As more debris melts down, a melt pool forms with the metallic material locating onto the oxidic layer. MPS method has advantages in predicting the mechanisms of melt behavior due to its capability in capturing the interfaces of multiphases and multicomponents.

Melt breakup can be divided into thermal breakup and hydraulic breakup according to the mechanism difference. Thermal breakup is caused by violent evaporation of water droplets that have been entrapped in the melt. By contrast, hydraulic breakup is caused by interface instabilities arising from the velocity difference and density gradient. Thermal breakup is difficult to simulate with MPS method because it involves violent evaporation of water. To the authors' knowledge, only Koshizuka et al. [25] conducted a tentative simulation on melt droplet fragmentation under the impingement of multiple water jets. The water evaporation process is taken into consideration. A critical density ratio of jet fluid to pool fluid, which affects the formation of melt filaments, is obtained. Koshizuka et al. [26] also simulated thermal fragmentation process of melt droplets during vapor explosion by assuming that multiple water jets impinge onto a melt droplet simultaneously. The jet speed is as high as 70 m/s. The numerical simulation agrees well with the experimental observation, as shown in Fig. 10.9.

Some simulations are conducted to investigate hydraulic breakup of melt jet. Ikeda et al. [27] conducted a preliminary analysis on the dynamics of water jet injecting into a pool of a denser liquid under nonboiling and isothermal conditions, which is a simulation of coolant injection into a melt pool. The density ratio of the denser fluid (Fluorinert) to water is 1.88. The penetration process is divided into two stages. In the first stage the jet is surrounded by an air pocket, and in the second stage the air pocket collapses and jet breakup is governed by Kelvin–Helmholtz and

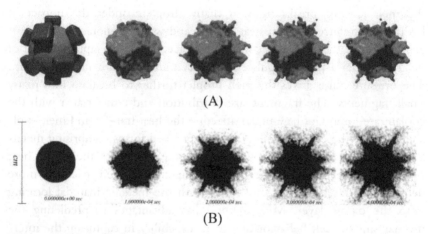

(A)

(B)

Figure 10.9 Melt droplet fragmentation in vapor explosions [26]. (A) Calculation result. (B) X-ray projection images obtained from experiment.

Rayleigh—Taylor instabilities. Shibata et al. [28] investigated jet breakup behavior by ignoring the surrounding continuous fluid. The effects of Weber and Froude numbers on jet breakup length and droplet size distribution are investigated, and the results agree well with the experimental data and correlations. Fig. 10.10 shows the jet breakup simulations under the conditions that the nozzle diameter is 0.013 m, the fluid density is 1000 kg/m^3, and the particle distance is 0.8125 mm.

In terms of hydraulic breakup of melt droplets, Nomura et al. [29] investigated melt droplet breakup behavior in a coolant pool during steam explosion. The boiling of water and the solidification of the melt droplet are not considered, but the effect of steam film surrounding the melt droplet is studied. It is found that the steam film can suppress the breakup of the melt droplet. In addition, the critical breakup Weber number of a melt droplet in the condition of steam film existing is approximately 50. The simulated breakup modes under the conditions of different Weber numbers agree well with that in the experiments. Duan et al. [30] investigated the deformation and breakup processes of a UO$_2$ droplet in a coolant pool with the Weber number at around a critical value and without the consideration of steam film. The melt droplet breakup process can be divided into two stages, that is, the stage of droplet stretching and thinning normal to flow direction and the stage of detach points appearing on the surface of droplets due to the growth of surface waves. Three breakup mechanisms are clarified, that is, the deformation mechanism of breakup,

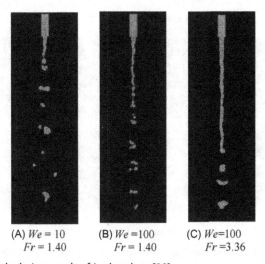

(A) $We = 10$ (B) $We = 100$ (C) $We = 100$
$Fr = 1.40$ $Fr = 1.40$ $Fr = 3.36$

Figure 10.10 Calculation result of jet breakup [28].

unstable growth of surface waves, and stripping fine fragmentations. The critical Weber number is 13 for uranium dioxide droplet breakup in water. The detached points presenting on the stretched melt droplet increase with the increase of Weber number. Li et al. [31,32] studied the size distribution and surface area of melt fragments during breakup process by considering melt droplets with and without steam film, as shown in Figs. 10.11 and 10.12. A uranium dioxide droplet with the diameter at 10 mm is impulsively accelerated by water flow. The melt hydraulic fragmentation is divided into two stages: the initial deformation and fragmentation stages governed by flow inertial force and the subsequent fragment coalescence phase under the influence of fluid surface tension. Through the investigation of the effect of steam film on melt fragmentation, it is found that the steam film rapidly moved

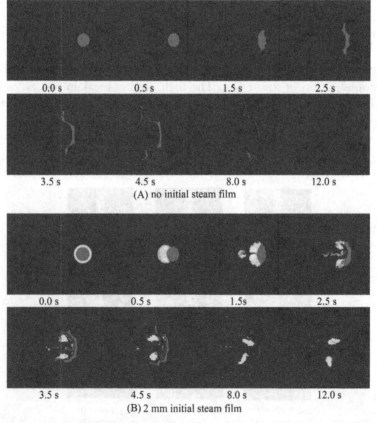

Figure 10.11 Melt droplet breakup behavior at $We = 1818$ [32]. (A) No initial stream film. (B) 2 mm initial stream film.

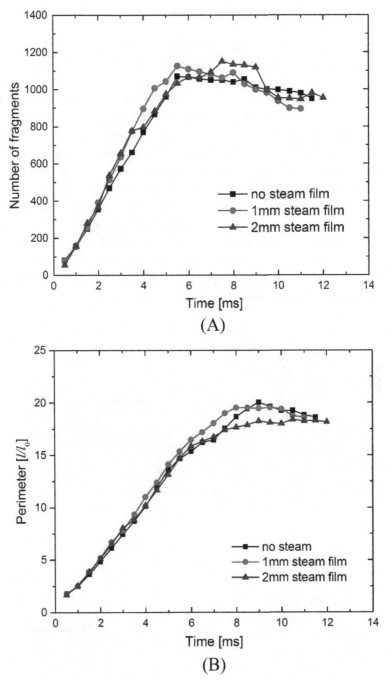

Figure 10.12 Fragment amount and melt—coolant contact perimeter during melt droplet breakup process at *We* = 2618 [32]. (A) Fragment amount. (B) Melt-coolant contact perimeter.

away under the effect of shear flow and it has an insignificant effect on fragment amount and surface area.

Park and Jeun [33], Park et al. [34], and Park et al. [35] simulated debris mixing and sedimentation processes after the melt fragment solidifies, which are related to debris bed formation and coolability. MPS method is coupled with a rigid body dynamic model to simulate the motion of debris. The developed method allows to identify key characteristics in debris sedimentation and debris bed formation processes, such as debris jet penetration, water leveling, and final debris configuration. The comparisons of two-dimensional (2D) and three-dimensional (3D) simulations indicate that 2D simulation is insufficient to predict debris sedimentation process because debris distribution depends on particle agglomeration and mixing prior to debris settling. Different from Park's debris modeling method, Muramoto et al. [36] modeled debris using the clusters of particles and evaluated fuel debris criticality in water by coupling MPS method with MVP code. Fig. 10.13 shows the fuel debris sedimentation process in water.

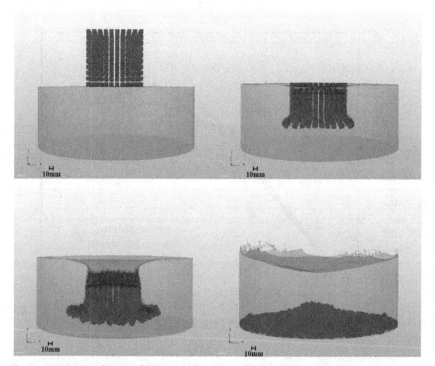

Figure 10.13 Simulation of falling 1000 cubes of debris into water [36].

The debris melting process was simulated by Duan et al. [37], in which the fluid—solid, solid—solid, and phase change models are coupled into MPS method. The PMS model and DEM are used to deal with the interactions of fluid—solid and solid—solid, respectively. To solve the problem of large computation cost of debris melting, a speedup algorithm is proposed, which calculates the conduction heat transfer and debris flow separately. The multiphase MPS method is applied to the simulation of debris melting in BWR lower head. The parameters adopted in the models, such as the spring stiffness in DEM model and relocation time and threshold melting ratio in the speedup algorithm, are sensitively investigated. Fig. 10.14 shows the results in the case of debris block radius at 0.073 m and the decay heat power at 0.7 MW/m^3. The analysis results indicate that the large debris blocks tend to result in severe local hot spots probably causing a lower breach point, but the small debris blocks probably lead to an upper breach point. The limitations in the study are also highlighted, where the convection heat transfer is underestimated and the melt drainage in the pore space is not well simulated in two dimensions. With the developed MPS method, Takahashi et al. [38] simulated debris melting scenarios that were properly present in the Fukushima Daiichi nuclear power plant. The analyses on the influences of debris size, initial debris distribution, and melt leakage around penetration structures are performed for units 2 and 3.

Following debris melting the phenomena of melt stratification and migration in debris bed are simulated by Li et al. [39—41]. Stratification behaviors of oxide and metallic core melt are investigated under different temperatures, viscosities, surface tensions, and boundary conditions. The results indicate that the crust formed at the oxide and metallic melt interface can impede the stratification progress to a large extent. Melt penetration in debris bed is terminated by crust, and the penetration distance increases with the increase of debris bed porosity and the increase of melt and debris temperatures. Fig. 10.15 presents melt stratification process when the metal and oxide melt simultaneously relocate into the lower head. Regarding the thermal erosion of the lower head wall by melt jet relocated from the core region, Li et al. [42,43] investigated the erosion rate of the lower head wall impinged by separate melt components. The fastest erosion rate presents in different conditions that have different melt and superheat temperatures. Fig. 10.16 shows the thermal erosion behavior of the lower head wall by UO$_2$ melt jet. Based on parametric studies, it is inferred that in the real severe accident the largest threat is from

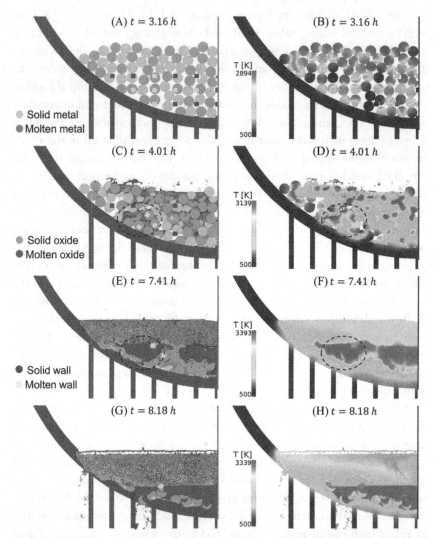

Figure 10.14 Debris remelting process in the case of the debris block radius at 0.073 m and the decay heat power at 0.7 MW/m³ (left: phase distributions, right: temperature distributions) [37].

metallic melt jet because thermally protective crust formed at the interface of oxide melt jet and the vessel wall prevents erosion progression.

The last in-vessel severe accident phenomenon is pressure vessel failure and melt discharge. There are different lower head failure mechanisms, such as melt leakage through penetration tubes, pressure vessel local melt

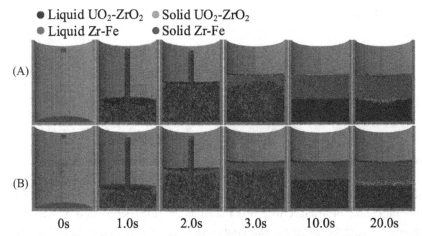

Figure 10.15 Stratification processes of oxide and metal melt under different viscosities (the initial melt temperatures in cases (A) and (B) are 2913K and 2900K, respectively) [41].

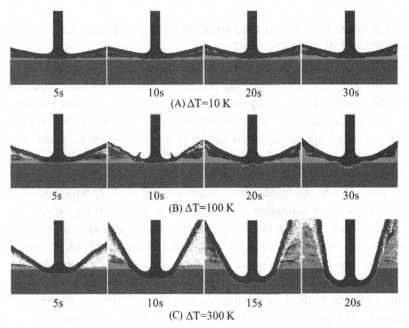

Figure 10.16 Thermal erosion of the lower head wall by UO_2 melt jet [43].

through, and global creep failure. Chen et al. [44] simulated melt penetration and solidification behavior in an instrument tube of RPV lower head. Melt flow resistance increases with the formation of melt crust. In

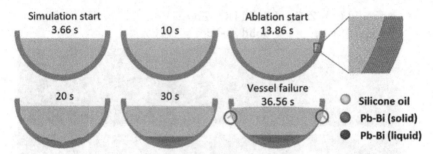

Figure 10.17 Phenomena of vessel wall melt through and melt discharge [45].

all the analyzed cases, melt does not penetrate through the tube due to the blockage of melt crust. Takahashi et al. [45] simulated vessel melt through using the simulant materials where the melt is represented by silicone oil and the vessel wall is made with lead-bismuth alloy. The vessel wall melt through and melt discharge scenario are presented in Fig. 10.17.

10.2.3 Melt spreading and molten corium–concrete interaction at containment

Following melt leakage into the containment, the melt spreads on the floor immediately. The spreading process involves free-surface flow, melt solidification, and potentially concrete ablation. MPS method has advantages in simulating these phenomena. Kawahara and Oka [46] simulated molten corium spreading behavior on ex-vessel floor and compared the melt front distance with that of FARO L26-S experiments. In the simulation the crust formation is not considered; that is, the solidified melt particles move with the melt flow. Yasumura et al. [47] investigated the influence of crust formation on corium spreading by immobilizing solid particles when its solid fraction reaches a threshold value and simultaneously it contacts with wall or fixed solid particles. The MPS method coupled with crust formation model is applied to the simulation of VULCANO VE-U7 experiments. Duan et al. [48] proposed a new algorithm of MPS method to simulate melt crust formation by increasing the viscosity. The problem of creeping velocity with the increase of viscosity is solved. With the new algorithm, the crust stress/strain can be calculated from the crust "physical creeping" based on the changes of relative positions of crust particles. The von-Mises yield criterion is adopted for fracture detection. Once a particle satisfies the fracture criterion, it is changed to a melt particle. As a result the melt can spread again beyond the crust breach [49,50]. The stop-and-go phenomena in the FARO spreading experiment are well predicted with the crust formation and fracture models, as shown in Fig. 10.18. In the further analysis of VULCANO VE-U7 spreading experiments by Jubaidah et al. [51], the effect of contact resistance at the melt—substrate interface on heat transfer and the influence of agitation effect of gas bubbles on melt spreading flow are modeled. Different melt spreading distances on ceramic and concrete substrates are predicted, and the mechanism of such differences is ascribed to the effect of gas bubbles generated during concrete decomposition.

Molten corium—concrete interaction (MCCI) can lead to ablation of concrete substrate, and as a result the containment integrity will be threatened. MCCI is an extremely complicated multiphase and multicomponent phenomenon involving concrete ablation and decomposition, gas release, melt crust formation, component migration, and so on. Many simulations with MPS method are performed by simplifying some phenomena. Koshizuka et al. [52] first applied MPS method to the simulation of an MCCI experiment named SWISS-2 by considering crust formation and melt natural convection. Li and Oka [53] analyzed MCCI experiments of SURC-2 and SURC-4, and the effect of Zr oxidation on concrete ablation was investigated. The results indicate that Zr oxidation significantly increases the ablation rate and the phenomena of crust formation and remelting are observed at the interface of melt and concrete. Crust can significantly slow down the concrete ablation rate. Later, the investigations on isotropic and anisotropic ablations of concrete are also conducted by carrying out numerical simulations of CCI-2 and CCI-3

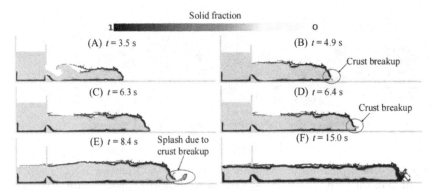

Figure 10.18 Melt spreading phenomenon with crust formation and breach (crust is black) [49].

experiments [54]. The interactions of fully oxidized PWR core melts with specially designed 2D limestone and siliceous concretes are analyzed. The influences of slag film, crust formation, gas generation, and aggregates on concrete ablation behavior are investigated. The ablation profile and the axial and lateral ablation rates agree well with the experimental data. The results also prove that aggregates are the cause of anisotropic ablation profiles in the interaction of melt with siliceous concrete. The representative ablation profiles in the simulation of CCI-3 experiment are shown in Fig. 10.19. In order to deeply understand the effect of aggregates on

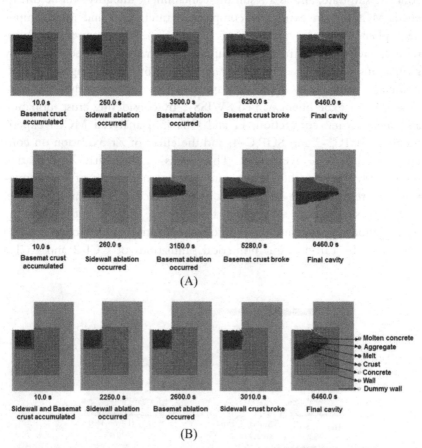

Figure 10.19 Representative moments of simulated MCCI process in CCI-3 experiment [54]. (A) Case 1 with aggregates, no sidewall crust model; Case 2 without aggregates, no sidewall crust model. (B) Case 3 with aggregates, with sidewall crust model.

concrete anisotropic ablation, 3D simulation of VULCANO VB-U7 experiment is conducted. The analysis indicates that the absent sidewall crust and aggregates at the bottom acting as a heat sink are the main causes of anisotropic ablation [55].

Chai et al. [56] analyzed CCI-2 experiment as well by coupling multiphysics models into MPS method. Crust thicknesses at axial and radial directions are predicted. The differences of concrete ablation profiles during interaction of melt with limestone and silica concretes are investigated

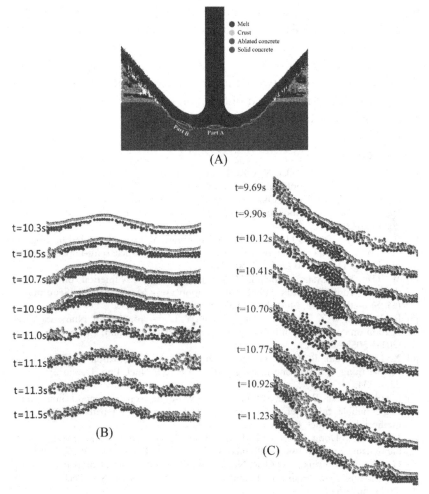

Figure 10.20 Crust evolution processes in the interaction of UO$_2$-ZrO$_2$ jet with concrete at melt temperature = 3000K and jet diameter = 20 mm [55]. (A) Erosion profile at 10.0 s. (B) Crust evolution at part A. (C) Crust evolution at part B.

[57]. In order to reduce the computational cost in 3D simulation of MCCI, Chen et al. [58] and Cai et al. [59] applied explicit algorithm and simulated the HECLA-3 and HECLA-4 experiments. The effect of crust formation at a low melt pouring rate on the anisotropic ablation is studied. Li et al. [60] investigated crust evolution in the interaction of molten corium jet with sacrificial material of the core catcher of European PWR. The results indicate that crust and molten concrete layers are present at the UO_2-ZrO_2 melt and concrete interface, whereas no crust forms in the interaction of Fe-Zr metallic melt with concrete. The crust experiences stabilization—fracture—reformation periodic process, and the molten concrete layer accumulates and escapes simultaneously. The crust evolution processes are presented in Fig. 10.20.

On the other aspects of MCCI, Dong et al. [61] investigated the effect of bubbles rising through two immiscible liquids on heat transfer. Fukuda et al. [62] studied the localized distribution of the metallic phase that has been observed in VULCANO VF-U1 experiment.

References

[1] H.Y. Yoon, S. Koshizuka, Y. Oka, Direct calculation of bubble growth, departure, and rise in nucleate pool boiling, Int. J. Multiph. Flow. 27 (2001) 277−298.

[2] S. Heo, S. Koshizuka, Y. Oka, Numerical analysis of boiling on high heat-flux and high subcooling condition using MPS-MAFL, Int. J. Heat Mass Transfer 45 (2002) 2633−2642.

[3] R. Chen, W. Tian, G.H. Su, S. Qiu, Y. Ishiwatari, Y. Oka, Numerical investigation on bubble dynamics during flow boiling using moving particle semi-implicit method, Nucl. Eng. Design 240 (2010) 3830−3840.

[4] W. Tian, Y. Ishiwatari, S. Ikejiri, M. Yamakawa, Y. Oka, Numerical computation of thermally controlled steam bubble condensation using Moving Particle Semi-implicit (MPS) method, Ann. Nucl. Energy 37 (2010) 5−15.

[5] R.H. Chen, W.X. Tian, G.H. Su, S.Z. Qiu, Y. Ishiwatari, Y. Oka, Numerical investigation on coalescence of bubble pairs rising in a stagnant liquid, Chem. Eng. Sci. 66 (2011) 5055−5063.

[6] X. Li, W. Tian, R. Chen, G. Su, S. Qiu, Numerical simulation on single Taylor bubble rising in LBE using moving particle method, Nucl. Eng. Design 256 (2013) 227−234.

[7] J. Zuo, W. Tian, R. Chen, S. Qiu, G. Su, Two-dimensional numerical simulation of single bubble rising behavior in liquid metal using moving particle semi-implicit method, Prog. Nucl. Energy 64 (2013) 31−40.

[8] R. Chen, C. Dong, K. Guo, W. Tian, S. Qiu, G. Su, Current achievements on bubble dynamics analysis using MPS method, Prog. Nucl. Energy 118 (2020) 103057.

[9] H. Xie, S. Koshizuka, Y. Oka, Numerical simulation of liquid drop deposition in annular-mist flow regime of boiling water reactor, J. Nucl. Sci. Technol. 41 (5) (2004) 569−578.

[10] G. Yun, Y. Ishiwatari, S. Ikejiri, Y. Oka, Numerical analysis of the onset of droplet entrainment in annular two-phase flow by hybrid method, Ann. Nucl. Energy 37 (2010) 230−240.

[11] N. Shirakawa, Y. Yamamoto, H. Horie, S. Tsunoyama, Analysis of flows around a BWR spacer by the two-fluid particle interaction method, J. Nucl. Sci. Technol. 39 (5) (2002) 572–581.

[12] N. Shirakawa, Y. Yamamoto, H. Horie, S. Tsunoyama, Analysis of subcooled boiling with the two-fluid particle interaction method, J. Nucl. Sci. Technol. 40 (3) (2003) 125–135.

[13] A. Ui, S. Ebata, F. Kasahara, T. Iribe, H. Kikura, M. Aritomi, Study on solid–liquid two-phase flow on PWR sump clogging issue, J. Nucl. Sci. Technol. 47 (9) (2010) 820–828.

[14] J. Xiong, S. Koshizuka, M. Sakai, Numerical analysis of droplet impingement using the moving particle semi-implicit method, J. Nucl. Sci. Technol. 47 (3) (2010) 314–321.

[15] J. Xiong, S. Koshizuka, M. Sakai, Investigation of droplet impingement onto wet walls based on simulation using particle method, J. Nucl. Sci. Technol. 48 (1) (2011) 145–153.

[16] Z. Wang, K. Shibata, S. Koshizuka, Verification and validation of explicit moving particle simulation method for application to internal flooding analysis in nuclear reactor building, J. Nucl. Sci. Technol. 55 (5) (2018) 461–477.

[17] Z. Wang, F. Hu, G. Duan, K. Shibata, S. Koshizuka, Numerical modeling of floating bodies transport for flooding analysis in nuclear reactor building, Nucl. Eng. Design 341 (2019) 390–405.

[18] K. Morita, S. Zhang, S. Koshizuka, Y. Tobita, H. Yamano, N. Shirakawa, et al., Detailed analyses of key phenomena in core disruptive accidents of sodium-cooled fast reactors by the COMPASS code, Nucl. Eng. Design 241 (2011) 4672–4681.

[19] A.P.A. Mustari, Y. Oka, Molten uranium eutectic interaction on iron-alloy by MPS method, Nucl. Eng. Design 278 (2014) 387–394.

[20] A.P.A. Mustari, Y. Oka, M. Furuya, W. Takeo, R. Chen, 3D simulation of eutectic interaction of Pb-Sn system using Moving Particle Semi-implicit (MPS) method, Ann. Nucl. Energy 81 (2015) 26–33.

[21] Y. Li, R. Chen, K. Guo, W. Tian, S. Qiu, G.H. Su, Numerical analysis of the dissolution of uranium dioxide by molten zircaloy using MPS method, Prog. Nucl. Energy 100 (2017) 1–10.

[22] Y. Li, W. Tian, R. Chen, G. Duan, S. Qiu, G.H. Su, Numerical investigation of oxidation and dissolution behavior in the fuel cladding using MPS-CV method, Nucl. Eng. Design 379 (2021) 111252.

[23] J. Wang, Q. Cai, R. Chen, X. Xiao, et al., Numerical analysis of melt migration and solidification behavior in LBR severe accident with MPS method, Nucl. Eng. Technol. 54 (1) (2022) 162–176.

[24] R. Chen, L. Chen, K. Guo, et al., Numerical analysis of the melt behavior in a fuel support piece of the BWR by MPS, Ann. Nucl. Energy 102 (2017) 422–439.

[25] S. Koshizuka, H. Ikeda, Y. Oka, Numerical analysis of fragmentation mechanisms in vapor explosions. JAERI Conference, 1997-011.

[26] S. Koshizuka, Y. Oka, Advanced analysis of complex thermal-hydraulic phenomena using particle method. GENES4/ANP2003, Sep. 15–19, 2003, Kyoto, Japan.

[27] H. Ikeda, S. Koshizuka, Y. Oka, H.S. Park, J. Sugimoto, Numerical analysis of jet injection behavior for fuel-coolant interaction using particle method, J. Nucl. Sci. Technol. 38 (3) (2001) 174–182.

[28] K. Shibata, S. Koshizuka, Y. Oka, T. Yamauchi, Numerical analysis of jet breakup behavior using particle method, in: GENES4/ANP, Kyoto, Japan, 2003.

[29] K. Nomura, S. Koshizuka, Y. Oka, H. Obata, Numerical analysis of droplet breakup behavior using particle method, J. Nucl. Sci. Technol. 38 (12) (2001) 1057–1064.

[30] R. Duan, S. Koshizuka, Y. Oka, Two-dimensional simulation of drop deformation and breakup at around the critical weber number, Nucl. Eng. Design 225 (2003) 37–48.
[31] G. Li, J. Zhang, Q. Yang, G. Duan, J. Yan, Numerical analysis of hydrodynamic fine fragmentation of corium melt drop during fuel-coolant interaction, Int. J. Heat Mass Transfer 137 (2019) 579–584.
[32] G. Li, P. Wen, H. Feng, J. Zhang, J. Yan, 2D MPS analysis of hydrodynamic fine fragmentation of melt drop with initial steam film during fuel–coolant interaction, Ann. Nucl. Energy 142 (2020) 107378.
[33] S. Park, Gyoodong Jeun, Coupling of rigid body dynamics and moving particle semi-implicit method for simulating isothermal multi-phase fluid interactions, Comput. Methods Appl. Mech. Eng. 200 (2011) 130–140.
[34] S. Park, H.S. Park, G. Jeun, B.J. Cho, Three-Dimensional modeling of debris mixing and sedimentation in severe accident using the moving particle semi-implicit method coupled with rigid body dynamics, Nucl. Technol. 181 (2013) 227–239.
[35] S. Park, H.S. Park, B.I. Jang, H.J. Kim, 3-D simulation of plunging jet penetration into a denser liquid pool by the RD-MPS method, Nucl. Eng. Design 299 (2016) 154–162.
[36] T. Muramoto, J. Nishiyama, T. Obara, Numerical analysis of criticality of fuel debris falling in water, Ann. Nucl. Energy 131 (2019) 112–122.
[37] G. Duan, A. Yamaji, Mikio Sakai, A multiphase MPS method coupling fluid–solid interaction/phase change models with application to debris remelting in reactor lower plenum, Ann. Nucl. Energy 166 (2022) 108697.
[38] N. Takahashi, G. Duan, A. Yamaji, X. Li, I. Sato, Development of MPS method and analytical approach for investigating RPV debris bed and lower head interaction in 1F Units-2 and 3, Nucl. Eng. Design 379 (2021) 111244.
[39] G. Li, Y. Oka, M. Furuya, M. Kondo, Experiments and MPS analysis of stratification behavior of two immiscible fluids, Nucl. Eng. Design 265 (2013) 210–221.
[40] G. Li, Y. Oka, M. Furuya, Experimental and numerical study of stratification and solidification/melting behaviors, Nucl. Eng. Design 272 (2014) 109–117.
[41] G. Li, P. Wen, H. Feng, J. Zhang, J. Yan, Study on melt stratification and migration in debris bed using the moving particle semi-implicit method, Nucl. Eng. Design 360 (2020) 110459.
[42] G. Li, M. Liu, G. Duan, D. Chong, J. Yan, Numerical investigation of erosion and heat transfer characteristics of molten jet impinging onto solid plate with MPS-LES method, Int. J. Heat Mass Transfer 99 (2016) 44–52.
[43] G. Li, M. Liu, J. Wang, D. Chong, J. Yan, Numerical study of thermal erosion behavior of RPV lower head wall impinged by molten corium jet with particle method, Int. J. Heat Mass Transfer 104 (2017) 1060–1068.
[44] R. Chen, Y. Oka, G. Li, T. Matsuura, Numerical investigation on melt freezing behaviour in a tube by MPS method, Nucl. Eng. Design 273 (2014) 440–448.
[45] N. Takahashi, G. Duan, M. Furuya, A. Yamaji, Analysis of hemispherical vessel ablation failure involving natural convection by MPS method with corrective matrix, Int. J. Adv. Nucl. Reactor Design Technol. 1 (2019) 19–29.
[46] T. Kawahara, Y. Oka, Ex-vessel molten core solidification behavior by moving particle semi-implicit method, J. Nucl. Sci. Technol. 49 (12) (2012) 1156–1164.
[47] Y. Yasumura, A. Yamaji, M. Furuya, Y. Ohishi, G. Duan, Investigation on influence of crust formation on VULCANO VE-U7 corium spreading with MPS method, Ann. Nucl. Energy 107 (2017) 119–127.
[48] G. Duan, A. Yamaji, S. Koshizuka, A novel multiphase MPS algorithm for modeling crust formation by highly viscous fluid for simulating corium spreading, Nucl. Eng. Design 343 (2019) 218–231.

[49] G. Duan, A. Yamaji, A novel approach for crust behaviors in corium spreading based on multiphase MPS method, in: 12th International Topical Meeting on Nuclear Reactor Thermal-Hydraulics, Operation and Safety (NUTHOS-12), Qingdao, China, October 14–18, 2018.

[50] Jubaidah, Y. Umazume, N. Takahashi, X. Li, G. Duan, A. Yamaji, 2D MPS method analysis of ECOKATS-V1 spreading with crust fracture model, Nucl. Eng. Design 379 (2021) 111251.

[51] Jubaidah, G. Duan, A. Yamaji, C. Journeau, L. Buffe, J.-F. Haquet, Investigation on corium spreading over ceramic and concrete substrates in VULCANO VE-U7 experiment with moving particle semi-implicit method, Ann. Nucl. Energy 141 (2020) 107266.

[52] S. Koshizuka, M. Sekine, Y. Oka, H. Obata, Numerical analysis of molten core-concrete interaction using MPS method, JAERI conference 2000-015.

[53] X. Li, Y. Oka, Numerical simulation of the SURC-2 and SURC-4 MCCI experiments by MPS method, Ann. Nucl. Energy 73 (2014) 46–52.

[54] X. Li, A. Yamaji, A numerical study of isotropic and anisotropic ablation in MCCI by MPS method, Prog. Nucl. Energy 90 (2016) 46–57.

[55] X. Li, A. Yamaji, Three-dimensional numerical study on the mechanism of anisotropic MCCI by improved MPS method, Nucl. Eng. Design 314 (2017) 207–216.

[56] P. Chai, M. Kondo, N. Erkan, K. Okamoto, Numerical simulation of 2D ablation profile in CCI-2 experiment by moving particle semi-implicit method, Nucl. Eng. Design 301 (2016) 15–23.

[57] P. Chai, M. Kondo, N. Erkan, K. Okamoto, Numerical simulation of MCCI based on MPS method with different types of concrete, Ann. Nucl. Energy 103 (2017) 227–237.

[58] R. Chen, Q. Cai, P. Zhang, Y. Li, K. Guo, W. Tian, et al., Three-dimensional numerical simulation of the HECLA-4 transient MCCI experiment by improved MPS method, Nucl. Eng. Design 347 (2019) 95–107.

[59] Q. Cai, D. Zhu, R. Chen, J. Deng, Y. Li, C. Dong, et al., Three-dimensional numerical study on the effect of sidewall crust thermal resistance on transient MCCI by improved MPS method, Ann. Nucl. Energy 144 (2020) 107525.

[60] G. Li, M. Liu, J. Wang, D. Chong, J. Yan, Crust behavior and erosion rate prediction of EPR sacrificial material impinged by core melt jet, Nucl. Eng. Design 314 (2017) 44–55.

[61] C. Dong, K. Guo, Q. Cai, R. Chen, W. Tian, S. Qiu, et al., Simulation on mass transfer at immiscible liquid interface entrained by single bubble using particle method, Nucl. Eng. Technol. 52 (2020) 1172–1179.

[62] T. Fukuda, A. Yamaji, X. Li, J.-F. Haquet, A. Boulin, Analysis of the localized metallic phase solidification in VULCANO VF-U1 with MPS method, Nucl. Eng. Design 385 (2021) 111537.

CHAPTER ELEVEN

Applications in ocean engineering

11.1 Hydrodynamics

11.1.1 Wave impact on a ship deck

Wave impact on a ship deck is an important phenomenon in ocean engineering, which may cause serious damages to the containers, hatch covers, and other structures on the deck. Moving particle semi-implicit (MPS) method has inherent advantages in simulating such shipping water phenomena involving free surfaces, large deformation, and complex boundaries. Shibata et al. [1−4] conducted a series of studies on the prediction of impact pressure on the deck. Fig. 11.1 shows the calculation domain. A solitary linear wave is initially placed in front of a bow model. The wave height and wave length are 0.1 and 1.37 m, respectively. The total particle number is 328,452 with an initial spacing of 1.0 cm. The surface tension is not considered in the simulation. Fig. 11.2 shows the shipping water behavior. The process of water flushing ship deck is simulated, and the predicted pressure on the deck has a trend agreement with the experimental data, as shown in Fig. 11.3.

The water wave profile and propagation process are simulated under the conditions of fixed wall boundaries, transparent boundaries, and highly

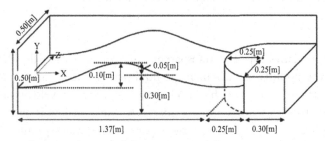

Figure 11.1 Calculation domain of shipping water [1].

Moving Particle Semi-implicit Method.
DOI: https://doi.org/10.1016/B978-0-443-13508-8.00011-1

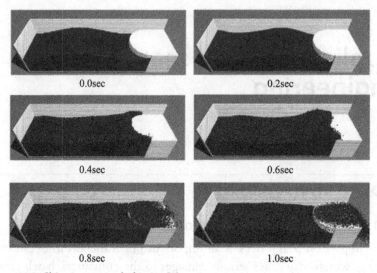

Figure 11.2 Shipping water behavior [1].

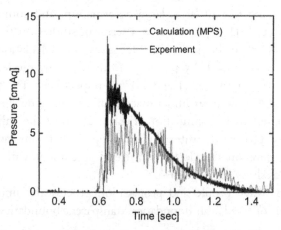

Figure 11.3 Time history of the impact pressure at a deck point [1].

viscous boundaries [3]. The transparent boundaries are actually the inflow and outflow boundaries, at which the fluid particles are dynamically generated and deleted. The Dirichlet boundary condition for velocity and pressure is set to the transparent boundary. A wave maker is produced by imposing wave velocity to the rigid panel. The wave profile is similar to Stokes wave, and the positions of wave amplitudes are predicted.

The incident waves can go through the transparent boundary, but they are overlapped by the reflected waves in the fixed boundary conditions.

In the condition of shipping water simulated with forced ship motion, the ship is modeled as a rigid body composed of rigid particles, and it is forced to move in regular head seas with the translational and rotational velocities [2]. In contrast with forced motion of the ship, a three-dimensional passive ship motion model is developed as well. The interaction between the rigid body and fluid is simulated by force and momentum consistency. A numerical wave tank is built with inflow and outflow boundaries to simulate the ship moving forward. The wave elevation and ship motion under different conditions are predicted [4]. Fig. 11.4 shows the shipping water behavior with passive ship motion model.

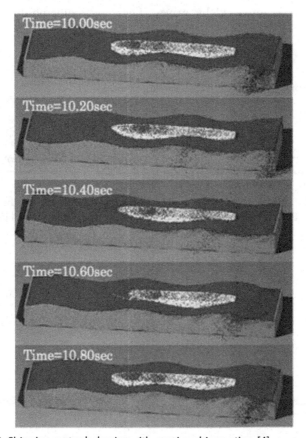

Figure 11.4 Shipping water behavior with passive ship motion [4].

In the wave impact simulation by Shibata et al. [1], the peak impact pressure is underestimated by about 50% and the pressure presents violent oscillation. Khayyer and Gotoh [5] investigated the shortcomings of MPS method, such as nonconservation of momentum and spurious pressure fluctuation, and proposed some improvements to MPS method for prediction of wave impact pressure. By revisiting the derivation processes of gradient model and pressure Poisson equation (PPE), new formulas of them are proposed. The improved MPS method is used to predict wave impact pressure. As shown in Fig. 11.5, a wave generator is placed on the left and the pressure on the vertical wall of right side is monitored. An effective suppression on pressure oscillation is realized by the improved MPS method, as shown in Fig. 11.6. Akimoto [6] simulated the flow around a high-speed planing boat to investigate the spray phenomenon. Sun et al. [7] investigated the violent hydroelastic problem when the ship hull structure is impacted by waves. The modal superposition model is

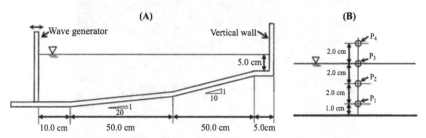

Figure 11.5 Simulation of wave impact: (A) schematic sketch of the numerical domain and (B) arrangement of pressure transducers at the vertical wall [5].

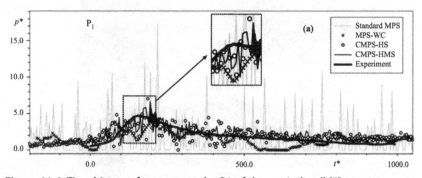

Figure 11.6 Time history of pressure at the P1 of the vertical wall [5].

coupled with MPS method to simulate the deformation of the flexible part and to handle the coupling effect between the rigid body and the flexible part.

11.1.2 Water flooding the damaged ship

Ship collision damage is a frequent accident, resulting in the fact that sea water floods into the ship. In the case of ship damage, the ship loses buoyancy and stability, and thus the survivability in flooding situations is an important problem that the ship designers and operators should consider. The conventional Euler-based grid methods have difficulties in dealing with flooding flow and complicated interaction with the damaged ship. Some attempts using the particle method to simulate the flooding flows into the damaged compartment of ships are conducted. Hashimoto et al. [8] proposed a hybrid method by calculating hydrodynamic forces of flooding flow with the MPS method and simulating intact parts with the potential flow theory. The hydrodynamic forces acting on the damaged ship are first calculated separately for the intact and damage parts and then summed together. The equations for modeling three degrees of freedom motion including sway, heave, and roll are expressed as

$$\left\{m_{22} + a_{22}\left(\omega_\phi\right)\right\}\frac{d^2x_2}{dt^2} + b_{22}\left(\omega_\phi\right)\frac{dx_2}{dt} + a_{24}\left(\omega_\phi\right)\frac{d^2x_4}{dt^2} + b_{24}\left(\omega_\phi\right)\frac{dx_4}{dt} = F_2^{MPS}$$

$$\left\{m_{33} + a_{33}\left(\omega_\phi\right)\right\}\frac{d^2x_3}{dt^2} + b_{33}\left(\omega_\phi\right)\frac{dx_3}{dt} + F_3^R(x_3, x_4) = F_3^{MPS}$$

$$\left\{m_{44} + a_{44}\left(\omega_\phi\right)\right\}\frac{d^2x_4}{dt^2} + b_{44}\left(\omega_\phi\right)\frac{dx_4}{dt} + a_{42}\left(\omega_\phi\right)\frac{d^2x_2}{dt^2} + b_{42}\left(\omega_\phi\right)\frac{dx_2}{dt} + F_4^R(x_3, x_4) = F_4^{MPS}$$

$$(11.1)$$

where m_{ij} is the mass or moment of inertia, a_{ij} is the added mass or added moment of inertia, x_i is the displacement or angle of motion, b_{ij} is the damping coefficient, F_i^R is the nonlinear restoring force, ω_ϕ is the natural angular frequency of roll, and F_4^{MPS} is the hydrodynamic force by MPS.

In order to make the MPS method applicable for the simulation of large-scale ships, the explicit algorithm together with graphics processing unit (GPU) acceleration technique is applied. The modified MPS method is validated by simulating forced roll motion of a flooded compartment with flooding water. The transient motion of the compartment and the flooding water flow are well predicted [8]. Fonfach et al. [9] simulated the sloshing flow in two compartments with an opening under the conditions of sway and roll motions and several water levels. The boundary condition is modified by extrapolating the pressure from inner wall particles to the dummy wall particles to model a slip and non-penetrable wall.

Zhang et al. [10] simulated the flooding of a freely floating damaged ship in two dimensions. Fig. 11.7 illustrates the analysis conditions including not only the holes located at different positions to represent broadside and bottom damage but also the internal horizontal and vertical nonpenetrating baffles. The effects of holes and the internal baffles on flooding amount and motion characteristics of the damaged ship are investigated, as shown in Fig. 11.8. The flooding amount from bottom damage is more than that from broadside damage. The vertical, horizontal, and roll motions are related to the distances of the hole to the ship mass center and to the calm water surface. Moreover, the vertical baffles have a greater influence on flooding than the horizontal baffles.

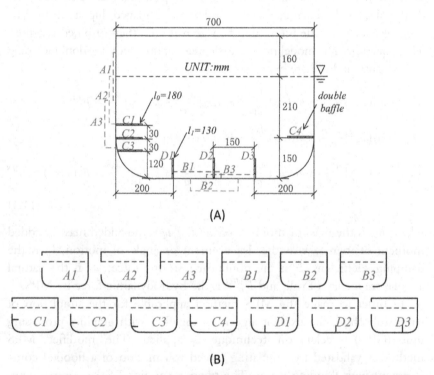

Figure 11.7 Model illustrations of the damaged ship [10]. (A) Details and dimensions of the damaged ship. (B) Simplified descriptions of cross section models of the damaged ship.

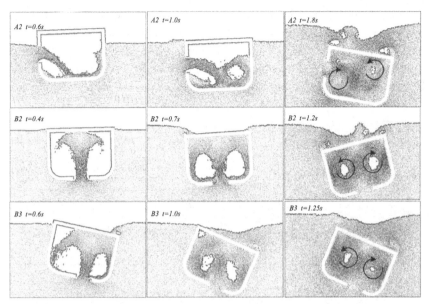

Figure 11.8 Flooding processes in ship broadside damage (model A2) and bottom damage (models B2 and B3) [10].

11.1.3 Sediment transport in waves

Landslides are complex natural phenomena that usually occur near mountains, oceans, bays, and reservoirs. Landslide-induced waves or tsunamis can lead to serious damage to the environment and social property. Meanwhile, the sediments, rocks, and clays carried by the landslide may change the terrain and the morphology of the coast. Landslides can be classified as nondeformable landslides and deformable landslides depending on the deformation of the sliding block, while based on the initial sliding position, landslides can also be classified as submerged landslides and unsubmerged landslides [11]. The continuum and discrete methods have been developed to model the granular flow of sediments, while the MPS method is used to model the wave flow. The landslide morphologies and the induced wave motion are investigated with the coupled method.

Shakibaeinia and Jin [12] introduced the Herschel–Bulkley rheology model to weakly compressible MPS method, forming a Lagrangian-based multiphase method, for the simulation of mobile-dam break problems. When the pores in the sediment bed are filled with water, the solid phase and pore water are considered as a mixture and defined as a non-Newtonian fluid. The rheology model is used to simulate the behavior of the non–Newtonian fluid, while the MPS method is used to model ambient water. The discontinuities of density and viscosity at the interface are dealt with average values. The multiphase MPS method is applied to mobile-bed dam break problems with different bed materials and initial bed shapes, as shown in Fig. 11.9. Fig. 11.10 shows the shape of the water

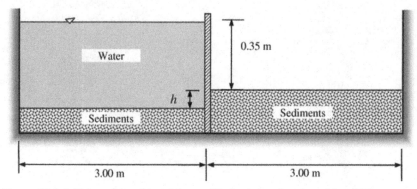

Figure 11.9 Initial configuration of the mobile-bed dam break problem [12].

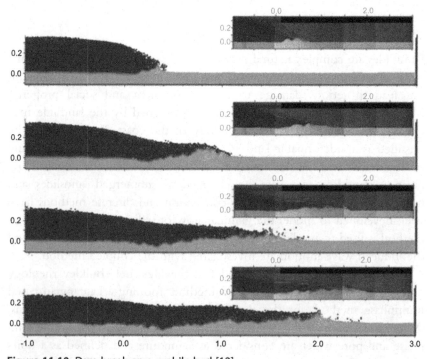

Figure 11.10 Dam break on a mobile bed [12].

and bed surface and the position of the leading edge, compared between the simulations and the experiments.

Fu and Jin [11] extended the multiphase MPS method by smoothing the material densities based on the particle numbers of different types in the effective radius and smoothing the viscosities using high-order Taylor expression. Then, both the nondeformable and deformable landslides in the conditions of initially submerged and unsubmerged are simulated. The slide-induced wave surfaces and sediment morphologies are well predicted. Fig. 11.11 shows the deformable landslide with/without rheology model and the effects of limiting viscosity in rheology model. Jin et al. [13] further investigated the landslide-induced waves with focus on wave propagation and the corresponding flow fields using the WC-MPS method. A nondeformable wedge block sliding along a slope is simulated in the conditions of the block released below the water surface, the block released immediately beneath the water surface, and subaerial release. Water surface variation and velocities around the sliding structure are compared between the numerical simulations and the experimental

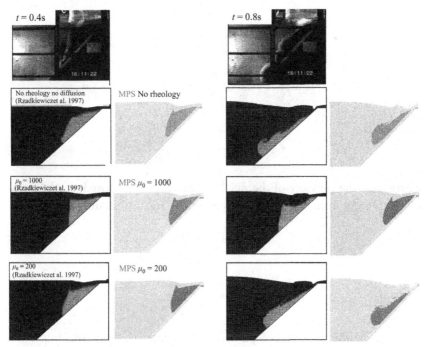

Figure 11.11 Deformable landslide with/without rheology model and the effects of limiting viscosity in rheology model [11].

results. Tajnesaie et al. [14] and Jandaghian et al. [15] also modeled the immersed granular flow that may be encountered in coastal engineering by a multiviscosity multidensity system. The mixture of solid grains and pore water is defined as a new non-Newtonian fluid phase, whereas the ambient water is a Newtonian fluid. A generalized rheological model using a regularized viscoinertial rheology is developed. With the developed method the gravity-driven granular flows in the subaerial and submerged landslides are analyzed.

Xu and Jin [16] developed a macroscopic MPS model to study the impact force on the face of a porous body that may be present in coastal engineering. The term "macroscopic" means that the detailed structures in the porous body are not modeled; instead the porosity is reflected in the modification process of MPS operators. When a particle moves from the nonporous domain to the porous domain with porosity φ, the particle distance is changed from l_0 to $l_0/\varphi^{0.5}$. In the porous domain, the PND, pressure gradient model, and Laplacian model are modified as

$$\langle n \rangle_i = \sum_{j \neq i} \frac{w_{ij}}{\varphi_j} \tag{11.2}$$

$$\langle \nabla p \rangle_i = \frac{d}{n^0} \sum_{j \neq i} \frac{p_j - \widehat{p}_i}{\varphi_j r_{ij}^2} w_{ij} \tag{11.3}$$

$$\langle \nabla^2 \psi \rangle_i = \frac{d}{n^0} \sum_{j \neq i} \frac{\psi_j - \psi_i}{\varphi_j r_{ij}^2} G_{ij} \tag{11.4}$$

$$G_{ij} = (3d - 2)w_{ij} + r_{ij} \frac{\partial w_{ij}}{\partial r_{ij}} \tag{11.5}$$

where φ_j is the porosity of the neighboring particle j. Moreover, the particle temporary motion and position correction are also modified as follows:

$$\mathbf{r}_i^* = \mathbf{r}_i^k + \frac{\Delta t \mathbf{u}_i^*}{\varphi_i} \tag{11.6}$$

$$\mathbf{r}_i^{k+1} = \mathbf{r}_i^k + \frac{\Delta t \mathbf{u}_i^{k+1}}{\varphi_i} \tag{11.7}$$

where φ_i is the porosity of particle i. In order to avoid the numerical instability caused by porosity discontinuity at the domain interface, a transition region is defined and within it the porosity is linearly interpolated.

Through the numerical simulations under different conditions, it is found that the porous materials with a larger porosity or a larger mean diameter are able to easily permit water penetration through the porous media, but the impact pressure at the interface and the reflected wave height are reduced [16].

A swash sediment transport process, which has an important influence to the beach topography change, was simulated by Harada et al. [17,18]. An improved MPS method by coupling the higher-order source term, error compensating part, higher-order Laplacian model, gradient correction, and dynamic stabilization to the differential operators (named MPS-HS-HL-ECS-GC-DS) is used to simulate fluid dynamics, while the discrete element method (DEM) is utilized for the simulation of sediment dynamics. The MPS-DEM coupled method allows to investigate the infiltration and exfiltration flow characteristics on a permeable beach face and the mechanism of ripple formation under swash zone. Harada et al. [19] further investigated the swash ripple formation mechanism in a mixed grain bed that is composed of coarse and fine-grained particles. The parameter of particle volume is used in the deduction process of MPS differential operators to consider the interaction of particles of different sizes. Two simulations are carried out for coastal morphodynamics analysis using the DEM-MPS method. One is the simulation under conditions where the DEM movable bed is fully submerged, with no pore water surface inside the DEM movable bed (see Fig. 11.12). The other one is the simulation under the conditions where the DEM movable bed includes a partially unsubmerged area, where the pore water surface exists in the DEM movable bed (see Fig. 11.13). The sorting process of coarse and fine solid particles can be well simulated, where coarse particles tend to accumulate in the crest of a ripple, while fine particles remain in the trough. The formation mechanism of a ripple is attributed to the selective transportation of solid particles; that is, the coarse particles move to and accumulate at the center part of a movable bed. With the motion of the particle bed the velocity of the small particles increases first, and then the velocity of large particles increases; consequently the constraints to the large particles from the surrounding small particles are reduced. The fluid drag force, pressure gradient force, and interparticle force are presented to explain particle transportation process. In the simulation of a solitary wave induced by a plunge of solid particles into water, the pore water level inside a solid-deposit layer and the wave height of a solitary wave are accurately estimated with the apparent volume change model. However, it should be

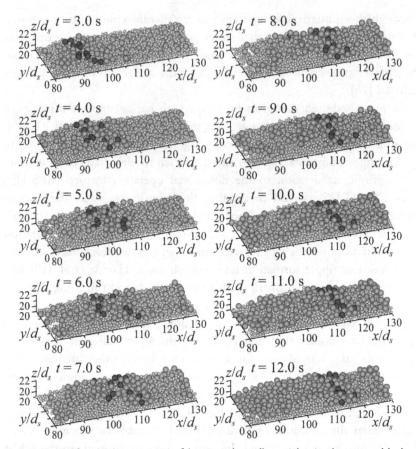

Figure 11.12 The sorting process of large and small particles in the moveable bed without the pore water surface [19].

noted that the resolution of fluid particles is not sufficiently high, and as a result the fluid passing through the gap among the solid particles cannot be solved accurately. Moreover, the 3D simulations are desirable to simulate the cohesiveness in the solid particles, but the computation cost will be very high and the pressure noise may present.

On the aspect of water wave flow without sediments, Ye et al. [20] simulated the dam-break wave propagation over dry and wet downstream beds to study the flow patterns, free surface, wavefront movement, and near-bottom shear stress. Wang et al. [21] simulated wave overtopping process on a sloping sea dike. The tsunami-like wave is produced by a moving wall. The wave deformation and hydraulic jump phenomenon of single and two successive solitary waves are investigated.

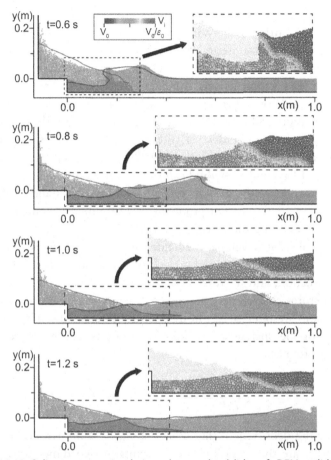

Figure 11.13 Solitary wave simulation due to landslide of DEM solid particles (the brown particles represent DEM solid particles, and the blue particles represent water) [19].

11.2 Fluid and elastic structure interactions

Sloshing in partially filled tanks is a phenomenon that can be observed frequently during liquid bulk cargo carriers operating on rough sea. The high nonlinear behavior of sloshing involving violent fluid motion and high impact pressure could potentially cause large deformation on the walls of tanks, loss of control of stability and maneuverability of the ship, particularly when the excitation associated with the motion of the ship is close to the natural sloshing frequency. The phenomenon, hence, is of great importance in assessing structural strength for designers of liquid cargo carriers [22]. With the development of computation technology, a number of numerical studies have been devoted to the simulations of FSI problems, especially the elastic structure. Most of numerical analyses for FSI problems are conducted in Eulerian framework with grids. Using grid-based method, it needs special boundary treatment for large deformation problems and an additional solver algorithm to regenerate grid configurations. However, these special treatments may degrade mass and momentum conservations at fluid and solid interfaces. The particle-based Lagrangian method can easily deal with fluid large deformation problems. Recently, smooth particle hydrodynamics and MPS methods have been applied to FSI problems by using fully Lagrangian particle method or combining them with grid method.

Song et al. [23] developed structure analysis model based on particle interaction models of MPS method, which can simulate elastic behaviors, large deformation, and fracture of structures. The equations of motion in elastic structures characterized by material density, Young's modulus, and Poisson's ratio are discretized through particle interactions. The governing equations for the analysis of an isotropic elastic structure in two dimensions are written as

$$\rho \frac{\partial^2 u_i}{\partial t^2} = \frac{\partial}{\partial x_j} \left[\lambda \varepsilon_{kk} \delta_{ij} + 2\mu \varepsilon_{ij} \right] \tag{11.8}$$

where ρ is the density, u_i is the displacement, δ_{ij} is the Kronecker's delta, and λ and μ are the Lame's elastic constants in two dimensions that are expressed by

$$\lambda = \frac{E\nu}{1 - \nu^2} \tag{11.9}$$

$$\mu = \frac{E}{2(1 + \nu)} \tag{11.10}$$

The first term at the right-hand side of Eq. (11.8) represents the pressure at the particle position, and the second term indicates the stress tensor between particles. Thus they are expressed as

$$p = -\lambda \varepsilon_{kk} \tag{11.11}$$

$$\sigma_{ij} = 2\mu \varepsilon_{ij} \tag{11.12}$$

Substituting Eqs. (11.11) and (11.12) into Eq. (11.8), Eq. (11.13) can be derived:

$$\rho \frac{\partial^2 u_i}{\partial t^2} = -\frac{\partial p}{\partial x_i} + \frac{\partial \sigma_{ij}}{\partial x_j} \qquad (11.13)$$

The strain tensor and rotation tensor between two particles are respectively calculated by

$$\varepsilon_{ij} = \frac{1}{2}\left[\frac{\partial u_j}{\partial x_i} + \frac{\partial u_i}{\partial x_j}\right] \qquad (11.14)$$

$$\omega_{ij} = \frac{1}{2}\left[\frac{\partial u_j}{\partial x_i} - \frac{\partial u_i}{\partial x_j}\right] \qquad (11.15)$$

The angular momentum of the particle is

$$I\frac{\partial^2 \theta}{\partial t^2} = M \qquad (11.16)$$

where θ is the rotation angle, which is the angle twice of the rotation tensor, and I is the moment of inertia.

The above governing equations are discretized by particle interaction models of MPS method. Song et al. [23] applied the models to the simulations of collision of two elastic rubber rings and chalk fracture. The numerical simulations agree well with the experimental phenomena. Yang and Zhang [24] used this structure model to FSI problems, where all the particles are first calculated using the fluid hydraulic equations, and then the velocities and displacements of solid particles are corrected by using the governing equations of elasticity. The advantages of this FSI model are that it is not necessary to do special treatment for the interface and the computation efficiency is high, but the results are not accurate enough.

Lee et al. [25] developed a fluid–shell structure interaction model by using particle and finite element coupled methods, where MPS method is used to analyze fluid flow and MITC4 shell element is used for FEM (finite element method) analysis of structures. The elements of the structure are defined by particle positions, that is, the connection of four particle mass centers forming one element in two dimensions. The outermost layer of structure particles participates in the pressure calculation of the fluid, and the pressure is directly applied to the structure analysis. Fig. 11.14 shows the treatment at the FSI interface. The fluid solver uses a semi-implicit algorithm, and the structure solver uses an explicit algorithm.

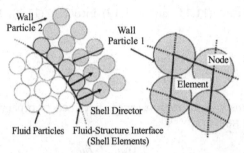

Figure 11.14 FSI (only wall particle 1 participates in the fluid pressure calculation and the structure element is defined by particle center connections) [25].

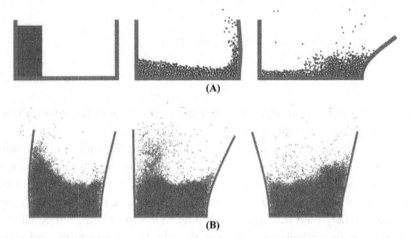

Figure 11.15 Simulations of fluid flow with the elastic wall [25]. (A) dam break with elastic wall. (B) Sloshing tank with elastic wall.

A partitioned coupling scheme is used between the fluid and structure solvers, and the Neumann–Dirichlet condition is applied to both the fluid and the structure. Even though the numerical results of the simulation of sloshing flow in elastic walled tanks are not stable, where fluid particles present unphysical splash, as shown in Fig. 11.15, it is of great significance for the application of MPS method to the FSI problems.

Hwang et al. [26] developed a fully Lagrangian particle method for the simulation of FSI problems by coupling the structure analysis model and the fluid analysis model with a partitioned-based FSI solver. Fig. 11.16 depicts the concept of a coupled system. In fact, it shows that a moving boundary is considered in the analysis of the fluid, which has the specific

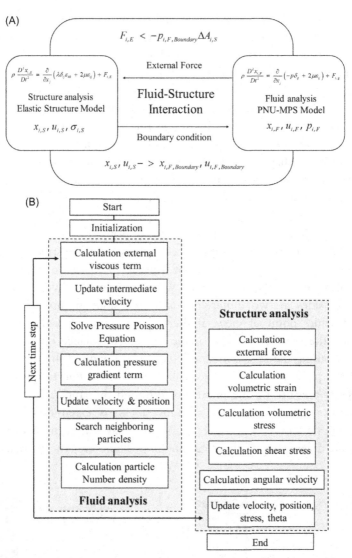

Figure 11.16 Coupling scheme of FSI model: (A) concept of the FSI coupling procedure and (B) algorithm of FSI simulation [26].

position and velocity fields. The pressure field of the fluid in the vicinity of the structure particles is considered as external force in the structure analysis. After structure analysis the positions and velocities of structure particles are used as the boundary conditions for the pressure calculation of fluid particles at the next time step. Therefore, the fluid boundary is updated in every time step by considering the instantaneous velocities and positions of structure particles. The updated pressure field of fluid particles is then used as the external force for the stress and velocity calculations in the next structure analysis. The geometric continuity and momentum conservation are achieved at the fluid–structure interface during these coupling processes. The fluid–structure coupling model is applied to the simulations of the elastic plate subjected to hydrodynamic pressure in the benchmark problem of dam break (see Fig. 11.17) and the violent sloshing flow with a hanging baffle. Khayyer et al. [27] improved FSI model by incorporating the high-order source term and error compensating parts of PPE and the dynamic stabilizer to the MPS-based fluid analysis model. The improved model is verified by the simulation of aluminum beam impacting on the water pool and the dam break with an elastic plate.

The above-mentioned coupling scheme for fluid and structure analysis may lead to discontinuous motion of structure particles. In the case of an elastic plate submerged in the static water, it bears the hydrostatic forces from two sides in two dimensions and four sides in three dimensions. From a statics point of view, if there exists no other active forces on the plate, the symmetrical hydrostatic forces will balance out each other. However, using the above-mentioned coupling method the surface particles of the plate may experience small displacements due to the action of hydrostatic pressure on the interface. Although the amplitude of the displacements is small, they bring about thickness extensional vibrations which will be increased gradually and finally may lead to numerical instability and divergence of the solution [28]. Therefore, the particle-based

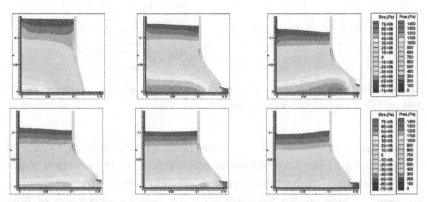

Figure 11.17 Simulation of the elastic plate subjected to hydrodynamic pressure in dam break [26].

FSI solver is improved by considering structural particles at the same elevation as one group, as shown in Fig. 11.18. As a result the force acting on a particle belonging to one group is calculated by dividing the total force by the total mass of the group; that is, the hydrostatic force is smoothed on the structure particles. A smoother force distribution on the elastic baffle is gained under the action of the fluid, compared with the previous coupling scheme [25], as presented in Fig. 11.19.

The MPS-FEM coupled solver for FSI problem is also proposed by Zhang and Wan [22]. The coupling strategy is similar to that used by Hwang et al. [26]. The fluid analysis is performed by considering the structure as the boundary, and the average pressure of the fluid particles is considered as the external force for the structure analysis. A small time step is used for fluid analysis according to the Courant–Friedrichs–Lewy (CFL) condition, whereas a much larger time step is used for the structure analysis. The structure particles are grouped according to their positions, which is like the scheme used by Hwang et al. [28]. The external forces

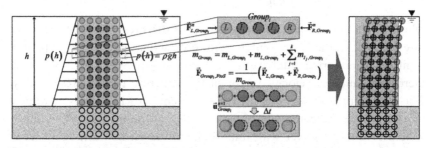

Figure 11.18 Smooth scheme of fluid–structure coupling force [28].

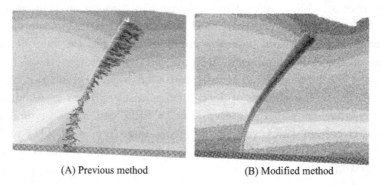

(A) Previous method   (B) Modified method

Figure 11.19 Force distribution on the elastic structure calculated by (A) the previous method [26] and (B) the improved method [28].

acting on the structure particles within the same group are averagely applied onto the structural FEM node. The structural particles within the same group move according to the linear and angular velocities. The MPS-FEM coupling solver is applied to the simulations of sloshing flow in liquid tanks with elastic lateral walls.

Khayyer et al. [29] further improved the consistency at the fluid—structure interface by revisiting the mathematical basis of MPS method. The accuracy and stability of fluid models are enhanced by using high-order discretization scheme. The multiresolution scheme is applied to reduce the computation cost. The structure model is set in the framework of Newtonian mechanics on the basis of conservations of linear and angular momenta with MPS-based discretization, as written below.

$$\left(\frac{D\mathbf{u}}{Dt}\right)^s = -\frac{1}{\rho^s}\nabla\cdot\boldsymbol{\sigma}^s + \mathbf{g} + \mathbf{a}^{FS} \tag{11.17}$$

$$\frac{D}{Dt}(I\boldsymbol{\omega})^s = \frac{D}{Dt}(\mathbf{r}\times m\mathbf{u})^s \tag{11.18}$$

where ρ^s is the density of the structure, $\boldsymbol{\sigma}^s$ is the stress tensor of the structure particle, and $\mathbf{a}^{FS}$ is the interaction force acting on a structure particle from its neighboring fluid particles; I and $\boldsymbol{\omega}$ represent the moment of inertia and angular velocity vector of the structure particle, respectively. The fluid—structure coupling between different particle scales is performed by considering the continuity of normal stress at the interface. The fluid—structure coupling solver with multiresolution is verified through a set of ocean engineering-related benchmark problems. Fig. 11.20 shows the hydroelastic slamming of a marine panel onto water.

Sun et al. [30] calculated the FSI problem through an iterative way, in which the Gauss—Seidel method with Aitken relaxation approach is adopted. The Neumann boundary condition is applied on the innermost layer of the solid boundary, and the pressure gradient is calculated between the particles on the innermost solid boundary and the nearest fluid particles. The stability of fluid calculation is improved by using error compensating parts to the source term of PPE, new types of solid and free-surface boundaries, and the PS technique. Sun et al. [31] further developed a coupled model of rigid body motion and modal superposition to study the interaction between violent water and flexible structures, in which the modal superposition model calculates the relatively small elastic deformations of the flexible structure. The mutual effect between

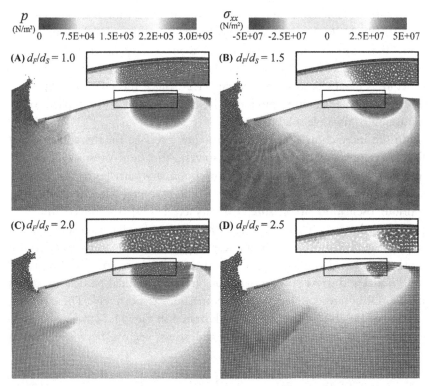

p
(N/m²)   0   7.5E+04 1.5E+05 2.2E+05 3.0E+05

σ_{xx}
(N/m²)   -5E+07 -2.5E+07   0   2.5E+07 5E+07

(A) $d_F/d_S = 1.0$ **(B)** $d_F/d_S = 1.5$

(C) $d_F/d_S = 2.0$ **(D)** $d_F/d_S = 2.5$

Figure 11.20 Stress (σ_{xx}) and pressure (p) fields at $t = 20.0$ ms with different diameter ratios reproduced by the enhanced multiresolution FSI solver in hydrodynamic slamming test of a marine panel in the case of $v = 4$ m/s [29].

rigid body motion and elastic deformation is considered, which affects the rigid body motion pattern. The fluid motion is solved with the modified MPS method that effectively suppresses pressure fluctuations. The proposed method is verified in the simulations of two nonlinear problems, that is, breaking water dam impacting a floating beam and flexible wedge slamming into water, with good agreements with the experimental results. Zhang et al. [32] investigated the interaction between waves and free rolling bodies with the modified MPS method including the wave-maker module and free roll motion module.

Sun et al. [33] proposed a FSI solver by coupling MPS method and DEM, and the DEM is improved by introducing a bonded-particle model to predict structure deformation. A deformation coefficient is introduced to evaluate the changes of structure deformation and displacement. In addition, the improved interface force calculation and multiple time-step algorithm are applied in the combination of MPS and DEM. The proposed MPS-DEM method with the bonded-particle model is applied to study the motion and deformation of a moored structure under wave impact, and the effect of Young's modulus on the stability of the deformable structure is discussed, as shown in Fig. 11.21. Since the DEM is adopted to describe the solid structure motion, it has the possibility to be applied to the simulation of multiple solid structures.

In the interaction at the fluid—structure interface for the conventional fully Lagrangian particle-based FSI solvers, the fluid pressure is usually imposed as the acceleration of structure boundary particles rather than the components of interest in stress tensors. This coupling scheme may lead to spurious oscillation of pressure. Falahaty et al. [34] introduced the stress points to the fully Lagrangian particle-based FSI solvers to suppress the unphysical oscillations. The nonlinear structure models using solely nodal integration based on moving least squares model (referred as MLS) and both nodal and stress point integrations (referred as dual particle dynamics model, DPD) are coupled with MPS method, respectively. The coupling scheme of DPD-MPS FSI solver is illustrated in Fig. 11.22, which considers consistent transfers of velocity and stress vectors between motion

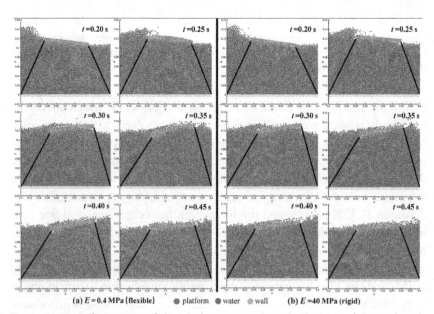

Figure 11.21 Deformations of the platform moored on the bottom of the tank with different Young's moduli [33].

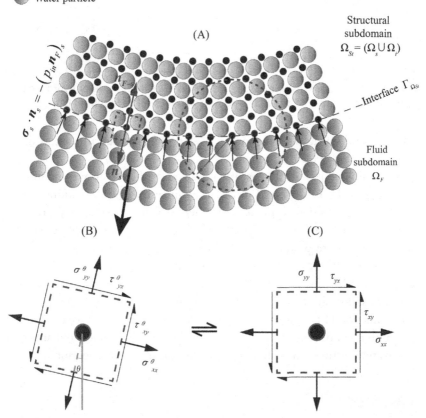

- Stress point
- Surface boundary stress point
- Structure motion particle
- Water particle

$$\sigma_s \cdot \boldsymbol{n}_s = -\frac{\sum_{j \in \Omega_F} p_j w(\boldsymbol{r}_j - \boldsymbol{r}_s)}{\sum_{j \in \Omega_F} w(\boldsymbol{r}_j - \boldsymbol{r}_s)} \boldsymbol{n}_{F-s}$$

(A)

Structural
subdomain
$\Omega_{St} = (\Omega_s \cup \Omega_I)$

Interface $\Gamma_{\Omega St}$

Fluid
subdomain
Ω_F

$\sigma_s \cdot \boldsymbol{n}_s = -(p_{int}\boldsymbol{n}_F)_s$

(B)   (C)

σ_{yy}^{θ} τ_{yx}^{θ} σ_{yy} τ_{yx}

τ_{xy}^{θ} τ_{xy}

σ_{xx}

σ_{xx}^{θ}

θ

Figure 11.22 Schematic diagram of DPD-MPS FSI solver. (A) Setup of the stress points at the inside boundaries of the structure, (B) surface stress point rotated to the direction of the surface normal vector, and (C) surface stress point rotated back to the direction of the global Cartesian coordinate [34].

particles and stress points. The coupled methods are comparatively validated in the benchmark tests of the hydrostatic water column on the elastic plate, dam break with an elastic gate, hydroelastic slamming of marine panels, and violent sloshing flows in tanks. The comparisons indicate that the combination of MPS method with DPD model, together with the modified interface boundary condition, shows remarkable improvements

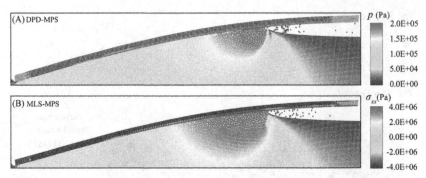

Figure 11.23 Comparisons of pressure and stress fields in the case of hydrodynamic slamming test of a marine panel by DPD-MPS and MLS-MPS FSI solvers [34].

in suppressing spurious oscillations related to zero-energy modes, as demonstrated by Fig. 11.23.

In summary, FSI problems are modeled with the above-mentioned solvers under the background of ocean engineering. The partitioned coupling scheme is widely applied, and different scales of time step are used for the fluid and structure calculations. The motion and displacement of elastic structure are predicted by considering the impact of the fluid flow. However, the work on the aspect of FSI is still preliminary, and the test cases for model verification are simple. Even so, the developed models provide a solid foundation for future studies on the stability, accuracy, and consistency of FSI simulations. FSI models are expected to be applied in complex conditions, such as structure fracture mechanics, dynamics of composite structures, and violent fluid flows in ocean engineering.

References

[1] K. Shibata, S. Koshizuka, Numerical analysis of shipping water impact on a deck using a particle method, Ocean Eng. 34 (2007) 585–593.

[2] K. Shibata, S. Koshizuka, K. Tanizawa, Three-dimensional numerical analysis of shipping water onto a moving ship using a particle method, J. Mar. Sci. Technol. 14 (2009) 214–227.

[3] K. Shibata, S. Koshizuka, M. Sakai, K. Tanizawa, Transparent boundary condition for simulating nonlinear water waves by a particle method, Ocean Eng. 38 (2011) 1839–1848.

[4] K. Shibata, S. Koshizuka, M. Sakai, K. Tanizawa, Lagrangian simulations of ship-wave interactions in rough seas, Ocean Eng. 42 (2012) 13–25.

[5] A. Khayyer, H. Gotoh, Modified moving particle semi-implicit methods for the prediction of 2D wave impact pressure, Coastal Eng. 56 (2009) 419–440.

[6] H. Akimoto, Numerical simulation of the flow around a planing body by MPS method, Ocean Eng. 64 (2013) 72–79.

[7] Z. Sun, G.Y. Zhang, Z. Zong, K. Djidjeli, J.T. Xing, Numerical analysis of violent hydroelastic problems based on a mixed MPS mode superposition method, Ocean Eng. 179 (2019) 285–297.

[8] H. Hashimoto, K. Kawamura, M. Sueyoshi, A numerical simulation method for transient behavior of damaged ships associated with flooding, Ocean Eng. 143 (2017) 282–294.

[9] J.M. Fonfach, T. Manderbacka, M.A.S. Neves, Numerical sloshing simulations: comparison between lagrangian and lumped mass models applied to two compartments with mass transfer, Ocean Eng. 114 (2016) 168–184.

[10] G. Zhang, J. Wu, Z. Sun, O.el Moctar, Z. Zong, Numerically simulated flooding of a freely-floating two-dimensional damaged ship section using an improved MPS method, Appl. Ocean Res. 101 (2020) 102207.

[11] L. Fu, Y.-C. Jin, Investigation of non-deformable and deformable landslides using meshfree method, Ocean Eng. 109 (2015) 192–206.

[12] A. Shakibaeinia, Y.-C. Jin, A mesh-free particle model for simulation of mobile-bed dam break, Adv. Water Resour. 34 (2011) 794–807.

[13] Y.-C. Jin, K. Guo, Y.-C. Tai, C.-H. Lu, Laboratory and numerical study of the flow field of subaqueous block sliding on a slope, Ocean Eng. 124 (2016) 371–383.

[14] M. Tajnesaie, A. Shakibaeinia, K. Hosseini, Meshfree particle numerical modelling of sub-aerial and submerged landslides, Comput. Fluids 172 (2018) 109–121.

[15] M. Jandaghian, A. Krimi, A. Shakibaeinia, Enhanced weakly-compressible MPS method for immersed granular flows, Adv. Water Resour. 152 (2021) 103908.

[16] T. Xu, Y.-C. Jin, Modeling impact pressure on the surface of porous structure by macroscopic mesh-free method, Ocean Eng. 182 (2019) 1–13.

[17] E. Harada, H. Gotoh, H. Ikari, A. Khayyer, Numerical simulation for sediment transport using MPS-DEM coupling model, Adv. Water Resour. 129 (2019) 354–364.

[18] E. Harada, H. Ikari, A. Khayyer, H. Gotoh, Numerical simulation for swash morphodynamics by DEM–MPS coupling model, Coastal Eng. J. 61 (1) (2019) 2–14.

[19] E. Harada, H. Ikari, T. Tazaki, H. Gotoh, Numerical simulation for coastal morphodynamics using DEM-MPS method, Appl. Ocean Res. 117 (2021) 102905.

[20] Y. Ye, T. Xu, D.Z. Zhu, Numerical analysis of dam-break waves propagating over dry and wet beds by the mesh-free method, Ocean Eng. 217 (2020) 107969.

[21] L. Wang, Q. Jiang, C. Zhang, Numerical simulation of solitary waves overtopping on a sloping sea dike using a particle method, Wave Motion 95 (2020) 102535.

[22] Y. Zhang, D. Wan, MPS-FEM coupled method for sloshing flows in an elastic tank, Ocean Eng. 152 (2018) 416–427.

[23] M.S. Song, S. Koshizuka, Y. Oka., Dynamic analysis of elastic solids by MPS method. GENES4/ANP2003, Sep. 15–19, 2003, Kyoto, Japan.

[24] C. Yang, H. Zhang, Numerical simulation of the interactions between fluid and structure in application of the MPS method assisted with the large eddy simulation method, Ocean Eng. 155 (2018) 55–64.

[25] C.J.K. Lee, H. Noguchi, S. Koshizuka, Fluid-shell structure interaction analysis by coupled particle and finite element method, Comput. Struct. 85 (2007) 688–697.

[26] S.-C. Hwang, A. Khayyer, H. Gotoh, J.-C. Park, Development of a fully Lagrangian MPS-based coupled method for simulation of fluid–structure interaction problems, J. Fluids Struct. 50 (2014) 497–511.

[27] A. Khayyer, H. Gotoh, J.-C. Park, S.-C. Hwang, T. Koga, An enhanced fully lagrangian coupled mps-based solver for fluid-structure interactions, 土木学会論文集B2(海岸工学) 71 (2) (2015) 883–888.

[28] S.-C. Hwang, J.-C. Park, H. Gotoh, A. Khayyer, K.-J. Kang, Numerical simulations of sloshing flows with elastic baffles by using a particle-based fluid–structure interaction analysis method, Ocean Eng. 118 (2016) 227–241.

[29] A. Khayyer, N. Tsuruta, Y. Shimizu, H. Gotoh, Multi-resolution MPS for incompressible fluid-elastic structure interactions in ocean engineering, Appl. Ocean Res. 82 (2019) 397−414.

[30] Z. Sun, K. Djidjeli, J.T. Xing, F. Cheng, Modified MPS method for the 2D fluid structure interaction problem with free surface, Comput. Fluids 122 (2015) 47−65.

[31] Z. Sun, K. Djidjeli, J.T. Xing, F. Cheng, Coupled MPS-modal superposition method for 2D nonlinear fluid-structure interaction problems with free surface, J. Fluids Struct 61 (2016) 295−323.

[32] Y. Zhang, Z. Tang, D. Wan, Numerical investigations of waves interacting with free rolling body by modified MPS method, Int. J. Comput. Methods 13 (4) (2016). 1641013-1-14.

[33] Y. Sun, G. Xi, Z. Sun, A fully Lagrangian method for fluid−structure interaction problems with deformable floating structure, J. Fluids Struct. 90 (2019) 379−395.

[34] H. Falahaty, A. Khayyer, H. Gotoh, Enhanced particle method with stress point integration for simulation of incompressible fluid-nonlinear elastic structure interaction, J. Fluids Struct. 81 (2018) 325−360.

CHAPTER TWELVE

Summary and outlook

The advantages of meshless particle method, characterized by easily capturing free surfaces, component tracking in multicomponent flow, and easily modeling solid—liquid phase transition, are complementary with the features of grid-based numerical method. Moving particle semi-implicit (MPS) method has got substantial improvements in past two decades, and the application areas have expanded from initial nuclear engineering to ocean, mechanical, and chemical engineering. Even though prominent advancements have been achieved in the aspects of stability, accuracy, boundary conditions, multiphase flow, and fluid—structure interaction (FSI), there are still many challenges in the practical applications of engineering problems. Therefore, further fundamental and comprehensive improvements are necessary to make MPS method a reliable and robust numerical tool in engineering analysis.

A critical defect of MPS method is its low accuracy and instability, which are about one order of magnitude lower than that of grid method. Exactly speaking, the standard MPS method only has a zero order of accuracy. Past studies have tried various approaches to improve its accuracy and stability, including artificial interparticle force, error-compensating parts in pressure Poisson equation (PPE) source term, PS scheme, and corrective matrixes for particle interaction models. Some improvements in accuracy and stability are gained with these treatments, but the conversation and convergence of numerical schemes are not strictly preserved because some of the models are not derived mathematically from fluid mechanics. A promising method to thoroughly resolve the low accuracy and instability problems of MPS method may be the least squares MPS method, which revisits fluid mechanics and Taylor series expansion. It does not need any artificial pressure stabilizer. Preliminary studies indicate that the stability and global high-order accuracy and consistency of least squares MPS method are demonstrated. In future, therefore, the studies should be focused on the derivation of flow-governing equations in a mathematical approach to preserve the conservation and convergence of numerical schemes. Like least squares MPS method, the calibration coefficient should be avoided.

Moreover, numerical schemes should be robust in the simulation of complex flows, not only in the limited conditions.

Another challenge of particle method is the modeling of multiphase flow with a large density ratio at the phase interface. Various approaches have been tested, including calculating two phases separately, density smoothing, and particle-grid hybrid method for gas—liquid flow. However, some of these treatments violate mass and momentum conservation laws, and the numerical schemes are not consistent at interfaces. In the future, research topics should focus on the conservations of volume, momentum, and energy at interfaces and the sharpness of interfaces to preserve original physical properties of materials. Considering the applications to solve multiphase flows in nuclear engineering, phase transition model is expected to be developed for evaporation and condensation simulations. In regard of solid—liquid flow, the MPS—discrete element coupling method is a promising method to calculate the interactions of solid—liquid and solid—solid. To make it more robust, stability enhancement techniques should be applied to the simulations of two-phase flow because the spurious hydraulic force caused by pressure fluctuation may lead to inaccurate motion of solid particulates. Moreover, the simulation of irregular solid particulates should be considered as well.

Development of turbulence models for strong turbulence flows are urgently needed because turbulence flow is a ubiquitous phenomenon in nuclear, ocean, and other relevant engineering. Several studies incorporate large eddy simulation models to smooth particle hydrodynamics and MPS particle methods for the simulation of low Reynolds turbulence flows. When this model is applied to high-intensity turbulence, the flow field cannot be predicted accurately, especially in the conditions with heat transfer. The concepts of time-averaged and spatially averaged turbulence models that have been widely applied in mesh-based method can serve as references to develop robust turbulence models for particle method. Moreover, heat transfer enhancement by turbulent vortexes should be taken into consideration in the solving process of energy conservation equations.

The boundaries are also troublesome problems for particle method. They are not obvious in the simple conditions. For some complex boundaries, however, it is difficult to arrange wall and dummy wall particles. Even though some advanced techniques, for example, polygon representation, are proposed, their implementation methods are complicated. Therefore, some simple boundary methods for complex boundaries are more practical. For the free-surface boundary, it is necessary to detect the free-surface particles accurately and make particle distribution uniform because the disturbance at the free surface is usually the cause of instability.

In regard to the applications in nuclear engineering, the simulations of some fundamental thermal hydraulic phenomena and melt behaviors have been conducted, but these simulations are simple and small-scale. It is expected to investigate thermal hydraulics considering more real reactor conditions, such as vaporization and condensation, and to analyze melt behavior using real melt properties and boundaries. Moreover, the capability of MPS method to conduct large-scale simulation is necessary and important. This capability should consider the reduction of MPS computation cost and the convenience to build particle configuration of complex geometry. On the other aspect, there is a gap between the current MPS technologies and the actual application; that is, the advanced techniques such as stability and accuracy improvements and multiresolution technique are not fully used in nuclear engineering.

Similarly, the MPS simulations on ocean engineering problems are carried out based on some simplifications. Many studies demonstrate the capability of MPS method on simulating the water wave propagation and FSIs. It is not difficult to find that the analysis conditions have large differences with the real conditions, for example, two-dimensional ship body and single plate representing a complex elastic structure. As a result, the six-degree-of-freedom motions of ship cannot be accurately simulated. Therefore, some high-efficiency computation techniques, such as parallel computation, multiresolution, and graphics processing unit acceleration, are expected to be applied to the simulations of ocean engineering problems. In terms of the simulations of FSIs, the accurate and consistent models for predicting structure deformation and fracture under violent fluid flow are expected.

MPS method is one representation of new generation computational fluid dynamic methods. In future, more advanced models will be continuously developed and more practical problems will be solved with this method. In addition, more researchers are encouraged to engage on the studies relevant to MPS method.

Index

W

Wall boundary, 71–73. *See also* Inflow and
    outflow boundaries
  conditions, 13–14
  distance-based polygon model, 77–78
  fixed boundary particle model, 74–77
  fixed wall (dummy) particle model,
    72–73
  integral-based polygon model, 78–80
  mirror particle model, 73–74
  particles, 145, 145*f*
  virtual particle-based polygon model,
    80–82
  wall boundary model, 72*f*
Wall particles, 72, 74
Wall weight function, 77
Water evaporation process, 198
Water flooding
  damaged ship, 219–220
    flooding processes in ship broadside
      damage and bottom damage, 221*f*
    model illustrations of damaged ship,
      220*f*

processes, 191
Water particles, 118
Water wave propagation, 215–217, 243
Wave impact, 215
  on ship deck, 215–219
  simulation of wave impact, 218*f*
Weight function, 7, 29, 62, 71, 145
  sketch for weight function in moving
    particle semi-implicit, 8*f*
Weighted least squares problem, 80–81
Wettability, 112

Y

Young's modulus, 227–228, 235–236,
    236*f*

Z

Zero-order gradient model, 54
Zero-velocity wall particles, 73
Zeroth-order accuracy, 121–122
Zirconium alloy, 193

Printed in the United States
by Baker & Taylor Publisher Services

Printed in the United States
by Baker & Taylor Publisher Services